Lecture Notes in Physics

New Series m: Monographs

Springer-Verlag Berlin Heidelberg GmbH

The Editorial Policy for Monographs

The series Lecture Notes in Physics reports new developments in physical research and teaching - quickly, informally, and at a high level. The type of material considered for publication in the New Series m includes monographs presenting original research or new angles in a classical field. The timeliness of a manuscript is more important than its form, which may be preliminary or tentative. Manuscripts should be reasonably self-contained. They will often present not only results of the author(s) but also related work by other people and will provide sufficient motivation, examples, and applications.

The manuscripts or a detailed description thereof should be submitted either to one of the series editors or to the managing editor. The proposal is then carefully refereed. A final decision concerning publication can often only be made on the basis of the complete manuscript, but otherwise the editors will try to make a preliminary decision as definite as they can on the basis of the available information.

Manuscripts should be no less than 100 and preferably no more than 400 pages in length. Final manuscripts should preferably be in English, or possibly in French or German. They should include a table of contents and an informative introduction accessible also to readers not particularly familiar with the topic treated. Authors are free to use the material in other publications. However, if extensive use is made elsewhere, the publisher should be informed. Authors receive jointly 50 complimentary copies of their book. They are entitled to purchase further copies of their book at a reduced rate. As a rule no reprints of individual contributions can be supplied. No royalty is paid on Lecture Notes in Physics volumes. Commitment to publish is made by letter of interest rather than by signing a formal contract. Springer-Verlag secures the copyright for each volume.

The Production Process

The books are hardbound, and quality paper appropriate to the needs of the author(s) is used. Publication time is about ten weeks. More than twenty years of experience guarantee authors the best possible service. To reach the goal of rapid publication at a low price the technique of photographic reproduction from a camera-ready manuscript was chosen. This process shifts the main responsibility for the technical quality considerably from the publisher to the author. We therefore urge all authors to observe very carefully our guidelines for the preparation of camera-ready manuscripts, which we will supply on request. This applies especially to the quality of figures and halftones submitted for publication. Figures should be submitted as originals or glossy prints, as very often Xerox copies are not suitable for reproduction. For the same reason, any writing within figures should not be smaller than 2.5 mm. It might be useful to look at some of the volumes already published or, especially if some atypical text is planned, to write to the Physics Editorial Department of Springer-Verlag direct. This avoids mistakes and time-consuming correspondence during the production period.

As a special service, we offer free of charge LATEX and TEX macro packages to format the text according to Springer-Verlag's quality requirements. We strongly recommend authors to make use of this offer, as the result will be a book of considerably improved technical quality.

Manuscripts not meeting the technical standard of the series will have to be returned for improvement.

For further information please contact Springer-Verlag, Physics Editorial Department II, Tiergartenstrasse 17, D-69121 Heidelberg, FRG.

Cherif Bendjaballah

Introduction to
Photon Communication

 Springer

Author

Cherif Bendjaballah
Laboratoire des Signaux et Systèmes
CNRS – ESE
Plateau de Moulon
F-91192 Gif-sur-Yvette Cedex, France

Library of Congress Cataloging-in-Publication Data

Bendjaballah, C. (Cherif), 1942-
    Introduction to photon communication / Cherif Bendjaballah.
        p.   cm. -- (Lecture notes in physics. New series m,
    Monographs ; m29)
    Includes bibliographical references.
    ISBN 978-3-662-14038-3   ISBN 978-3-540-49206-1 (eBook)
    DOI 10.1007/978-3-540-49206-1
    1. Optical communications. 2. Squeezed light. 3. Quantum optics.
I. Title. II. Series.
TK5103.59.B46   1995
621.382'7--dc20
                                                                95-20209
                                                                CIP

ISBN 978-3-662-14038-3

© Springer-Verlag Berlin Heidelberg 1995
Originally published by Springer-Verlag Berlin Heidelberg New York in 1995
Softcover reprint of the hardcover 1st edition 1995

Typesetting: Camera-ready by the author
SPIN: 10127333   55/3142-543210 - Printed on acid-free paper

Preface

In recent years, progress in the generation of squeezed states of light, mainly characterized by a reduced noise property, has stimulated important work in relation to their potential use to improve the sensitivity of optical communication systems. To perform below the quantum limit, it has been suggested that squeezing techniques could be employed. In order to discuss such applications, these notes will concentrate on the detection and transmission of information using quantum devices. At this point of the research and because most of the results have been published in many books and journals, we believe that a survey as an introduction to the field would be useful.

The aim of these lecture notes is mainly to present an introduction to the results which are essential for both physicists and engineers who are interested in present and future developments in optical communications using light sources with a very low level of power. Although the major emphasis is on theory, practical details are included occasionally in order to show the possible implementation of quantum optical communication systems using existing devices. A review of the basic ideas and concepts is presented in a concise manner, and the theory is given in a simple mathematical formalism. The most significant results have been discussed, as much as possible, according to their practical interest. The final form is necessarily incomplete because of some limited understanding and possibly a restricted selection of the topics.

The material is organized in four sections. Each section is almost self-contained and is followed by a bibliography for further reading. Both points of view, classical and quantum, are presented whenever it is possible. In the last section, additional comments on the phase operator have been added to cover some recent developments.

The advice and encouragement of Professor W. Beiglböck, Editor of Springer-Verlag, Heidelberg, have been most helpful. His interest in this work is gratefully acknowledged.

Gif-sur-Yvette
March 1995   Ch. Bendjaballah

Contents

Introduction 1

I. Elements of the Mathematical Formulation
   of the Quantum Theory 18

    I.1   States, Operators in Quantum Communication 18
    I.2   Representations 30
    I.3   Coherent Receiver Operators 47
    I.4   Measurement Processes 53
    I.5   References 60

II. Performance Criteria: Detection, Information 63

    II.1   Classical Formulation 63
    II.2   Quantum Expressions 84
    II.3   References 111

III. Direct Detection Processing 114

    III.1   Random Point Processes Theory 115
    III.2   Detection Performance 143
    III.3   Information Performance 156
    III.4   References 173

IV. Phase Operator 177

    IV.1   Some Models 178
    IV.2   References 191

Conclusion 193

Introduction

Since the important formula of Shannon showing that, for an ideal band–limited Gaussian channel, the channel capacity is proportional to the channel bandwidth, many engineers have tried to take advantage of optics to develop optical communication systems. Because of the lack of high quality optical sources and detectors, practical users of optical communication systems have until recently made little progress. In the last few years, however, technological progress in the generation of laser light of high quality frequency accuracy and stability, as well as in the production of low loss optical fibers, has stimulated important research work in the design of systems for optical communication.

In the usual communication systems, such as the radio frequency systems, the signal information to be sent propagates through a channel and is detected by a receiver in an optimal way in accordance with some detection or (and) information criteria. All precise and noise–free physical operations that are requested by the mathematical calculations of the optimization can be performed by the receiver. The only limitations are due to the unavoidable noise and are included in the description of the channel.

At optical frequencies, these assumptions are no longer valid because of the laws of the quantum mechanics. For instance, the exact measurement of the two quadratures of the optical field error is impossible because of the uncertainty principle.
Another consequence of the quantum mechanics limitations, is that if one measures only one quadrature component, the result is a time continuous random function even for completely specified light field and receiver properties.

Therefore, a fully quantum mechanical description of optical communication systems seems necessary. Fortunately, in some situations, a semi–classical modelization, treating the input field quantum mechanically and considering the channel and the detector from a classical point of view, is efficient. Thus, a direct detection processing using a photomultiplier receiving a light beam (e.g. laser or natural light) is sufficient and the performances derived on statistical detection as well as information performance, show the superiority of such a system over a completely classical system. It remains, however, to compare such performances with the fundamental limits called *optimal performances* that are derived from the quantum theory.

To evaluate these optimal performances, according to some detection (or information) criteria, the system being subject to some constraints (energy, power, bandwidth, ...), one may choose to restrict the class of signals to some typical states (e.g. coherent, squeezed) and seek the optimal quantum receiver which obeys the physical laws of measurements. Although this approach does not imply that the resulting optimal quantum operator represents a physical apparatus, it has been shown that the minimized probability of error in detection is achieved with existing devices. One may also choose to fix the receiver and optimize the system with respect to the signals and seek the optimal quantum state.

This approach is not appropriate for at least two reasons. First, there is no theoretical restriction in the specification of states. Secondly, there is no guarantee that the optimal state can be pratically produced.

To summarize, we can state that although the semi–classical formulation is appropriate in many practical situations, a rigorous theory requires a quantum mechanical treatment of the light source, the channel and the receiver. This is because, although there is some analogy between radioelectricity and optics, quantum mechanics is essential when dealing with photons. This is also because the performances attained by practical systems must be compared with ultimate limits derived from the quantum theory.

This quantum theory of detection required to deal with detection of optical signals at very low level is now well established at least in some ideal situations. It has recently been extended to take into account the properties of the recently generated non–classical states of the light field. The fundamental limits of various functions of statistical interest in the detection and transmission of information have indeed been calculated. Despite the fact that these results are obtained only for special cases, they are important because they permit a comparison with the performances attained in practice.

These performances obviously depend on the type of system of optical communication under study but also on the statistical properties of the noise. However, it has recently been proved that the most important problem in the optical transmission of information is the message encoding instead of the noise. This important conclusion has stimulated many works on the subject because it became necessary to develop a quantum theory of information and encoding. Such a work appears very difficult because of the mathematics involved but also because there is no particular restriction on the encoding properties of the states (source).

Besides these questions of mathematical aspect, problems of technological aspect appear in the three elements of the system: source–channel –receiver. To simplify the presentation, the channel (transmitter) is considered as a macroscopic object and the optical components are supposed perfect: all the degradations due to surface imperfections, finite aperture, misalignment, ... are not considered. Essentially the practical problems are the frequency (polarization, phase) stability and the control power of the lasers, the design of modulators, the adapted antennas, and the efficiency of the detectors.

Although, the fundamental limitations occur only when optical signals are of a very low level that is to say in a very limited number of cases, there are some practical situations in which quantum optical communication is greatly needed notably in long distance free space communication, quantum cryptography, or for quantum computer applications as suggested by Bennett and col. For example, the quantum cryptography as was recently practically demonstrated using two interferometers of Mach–Zehnder type to transmit and to detect events corresponding to photons of the same state, requires highly stabilized phase interferometers, very low noise modulators and very efficient detectors operating with a very weak power source of photons. Although requiring optical components of very high quality, the technique using polarization approaches could be easier to implement.

Now, even if the technological problems are overcome, quantum optical communication may prove too expensive and too complex to be competitive with classical optical communication systems. As a matter of fact, complicated problems will make it very difficult to implement of any of these schemes especially for producing stable, reproducible special states of light. This is also because it is uncertain that the quantum receiver would correspond to a realistic device.

Consider first a typical system of communication. Such a system can be sketched as follows in order to define parts carrying out three separate functions. The scheme is a simplified one because in general, the system must be treated as a whole.

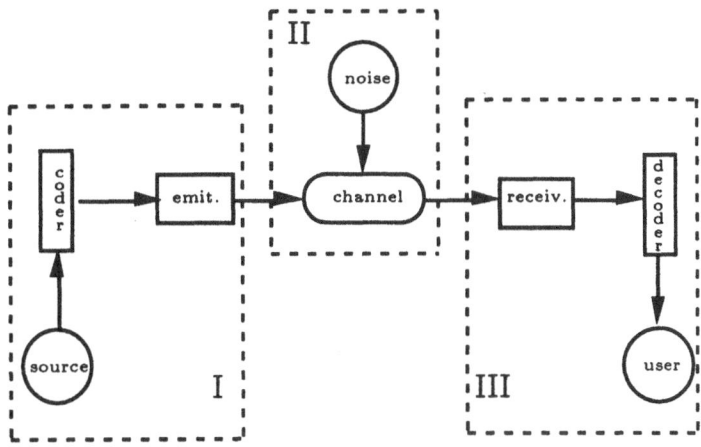

Fig. 0.1. Diagram of a typical communication system.

The scheme for such a system of communication can be divided into the three basic groups of elements: I. source (e.g. semiconductor laser), modulator (coder) conveying messages to be transmited, II. guide, e.g. optical fiber or free–space (channel), and III. receiver (e.g. photodiode) followed by a processor (decoder). The system analyzed here is simplified such as to be linear, memoryless, without feedback, without side–information (information link between user and source),$\cdots$ All parts of the system are obviously required to provide high speed, high reliability (parameter to characterize the reproducibility of the data in the same practical conditions), and high accuracy (parameter to characterize the precision of the measurement) performances for a undistorted signal transmission. This goal can indeed be attained as recent research proves that reliable information can be transmitted at a rate of $\geq 5Gb/s$ over very long distances ($\sim 10^4$kms).

To reach such a performance and to avoid the difficulties and costs due to the opto–electronic repeaters, the technique employed is of optical solitons such that the system includes a semiconductor diode laser of wavelength $\lambda_0 = 1.550\mu m$ and a power of $\sim 5mW$, a series of erbium–doped optical fibers as amplifiers and a photo-diode as a receiver. The sources of noise which come mainly from the amplified spontaneous emission, will limit the sensitivity (parameter that characterizes the possibility to detect the amplitude or the power of a signal) of performance to a great extent.

To briefly describe the physical process yielding soliton pulses, let us consider a light pulse, of high intensity and very short in time ($\sim 20-40ps$), that propagates along a fiber. The index of refraction of this fiber is a nonlinear function of the frequency and the light intensity $n(\omega, e) = n(\omega) + \chi e^* e$ due to the group velocity dispersion and to the Kerr effect respectively.

Around the central frequency $\omega_0 = \frac{c}{2\pi\lambda_0}$, the wave number is expanded up to the second order, denoting $k_0' = \frac{dk}{d\omega}|_{\omega_0}$ and $k_0'' = \frac{d^2 k}{d\omega^2}|_{\omega_0}$ the first two derivatives with respect to the frequency.

The electric field envelope $e(z, t)$, assumed to be slowly varying in time, describes the propagation along the z-axis and obeys the nonlinear Schrödinger equation $i\dfrac{\partial e}{\partial z} - \dfrac{k_0''}{2}\dfrac{\partial^2 e}{\partial \tau^2} + \chi|e|^2 + i\beta e = 0$ where β is a coefficient taking into account the unavoidable light attenuation, and where we set $\tau = t - k_0' z$.

When the effects compensate ($k_0''\chi < 0$), a stable multiple soliton pulses occur. These pulses can be transmitted without a significant loss and without any interference. Such properties are of course remarkable for very long distance optical communications purposes.

Here we do not consider such systems because we are mainly interested in quantum aspects of optical communications. We just focus on some physical concepts in the generation of optical source and analyze the receiver configuration in detail.

The light sources of interest here are essentially the conventional laser especially of a semiconductor type with a direct application to communications. The type of memoryless source modelised by a random vector $\overrightarrow{X}$ to deliver a set of messages taken in a finite alphabet $\mathcal{D}$ of cardinality d defined by a probability distribution $p(X_i = x_i)$ which is considered as stationary $\forall l, d, \ p(X_1 = x_1, ..., X_d = x_d) = p(X_{l+1} = x_1, ..., X_{l+d} = x_d)$ and time–invariant $\left(p(x_{d+1}|x_d) \text{ does not depend on } d\right)$. A simple model of a source with dependent messages is a Markov sequence for which $p(x_1, ..., x_d) = p(x_1)p(x_2|x_1)...p(x_d|x_{d-1})$. Notice, however, that it might also be possible to take into account a possible memory effect of the source by representing such a source as a combination of a memoryless source and a colored noisy channel.

One of the light sources for optical communications is mathematically modelized by a *coherent state*. To very shortly characterize such light field, recall that a light field can be written as a complex sum of the two fluctuating quadrature components $\overrightarrow{E_c}, \overrightarrow{E_s}$ of variances σ_c, σ_s respectively. These components obey the Heisenberg inequality type $\sigma_c \sigma_s \geq \delta^2$. In the case of $\sigma_c = \sigma_s = \delta$ such a field defines a coherent state. Obviously, there exist other fields generated in states such that $\sigma_c < \delta$ and $\sigma_s \geq \frac{\delta}{\sigma_c}$, states for which the fluctuations of one component are less than the *standard deviation* (variance $= \delta$) and verifying Heisenberg inequality. Such states are called *squeezed states*.

If a light field is in coherent state, the fluctuations for both variables cannot be modified. If a light field is in squeezed state, the fluctuations of one component can be controled. This is a very interesting feature of such a field. Thus, any measurement of light in the direction fixed by $\overrightarrow{E_c}$ will yield data of smaller fluctuations than the standard deviation. It is therefore clear that the construction of such new source is of major importance in quantum communication.

Here we shall just enumerate some physical processes leading to the generation of such squeezed states: essentially it is a parametric amplification process based on the nonlinearity of a crystal.

- Four waves mixing: the effect can be stimulated using the coefficient $\chi^{(3)}$ of an optical fiber.
- Three waves mixing: the effect can be stimulated using the coefficient $\chi^{(2)}$ of a crystal (e.g. $LiNbO_3$). Whenever the photon noise is very weak, the squeezing effect can be observed.
- Current control: the emitted photons are controlled by a periodical emission of electrons.

In the systems of interest here as depicted in Fig. 0.1, the transmission channel is mainly characterized by the properties of the modulation devices. Here again, we limit ourselves to memoryless channels so that the message undergoes a no memory modulation. For example, a modulator of the characteristic $m(t)$, produces for $t \in [0, T]$ the complex amplitude of the classical field such that $E \propto \mathcal{E}(\frac{z}{c} - t, m(\frac{z}{c} - t))$ does not depend on another time $\vartheta \in [0, T]$.

Although the light can be modulated from the pumping current, it is sometimes necessary to use an external modulator. These devices are of three types depending on the physical properties of the devices: electro–optical (the electrical field changes the index of refraction), acoustic–optical (the frequency of the sound field changes the angle of diffraction), magneto–optical (rotation of the polarization by Faraday effect). We consider the modulation systems that use a current modulator. Only *classical modulators* whose mode of operation is described by a scalar function of time are considered here. Spatial light modulators that are the key components for optoelectronic systems of image communications are not discussed in the context of these notes. However, there are many other modulation schemes of interest based on amplitude modulation, frequency modulation, phase modulation, and their digital correspondences (see below OOK, PPM): amplitude shift keying (ASK), frequency shift keying (FSK), and phase shift keying (PSK). These schemes can be realized by means of an external modulator for output light from a stable continuous laser by current technology. As pointed out, the modulation formats can be essentially of analog or digital types.

In the analog modulation scheme, the carrier can be time–modulated by a signal in one of various ways: amplitude, frequency, phase. In the digital method, the signal is sampled at equal regular time intervals. These samples are then quantized and coded. In general, however, for technological reasons, it is more interesting to deal with the digital system.

In such a system (M–ary), the signals at the modulator input take one of the M possible signals that may be transmitted during each T seconds. The communication system is called binary for ($M = 2$). More complex modulation schemes such as pulse–amplitude, pulse–width, pulse coded modulations, ... are also sometimes employed, but we will here limit ourselves to only two: on-off keying (OOK) and pulse position modulation (PPM).

In the case of OOK, the source of messages sends one of two possible symbols denoted "1" (presence of the message in a noisy signal), or "0" (absence of the message). The user decides after processing which symbol was sent (Fig. 0.2a).

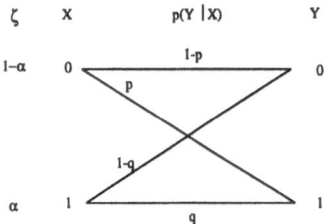

Fig. 0.2a. X–channel

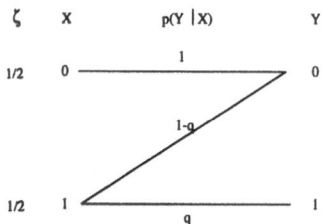

Fig. 0.2b. Z–channel

For the direct detection scheme that we will be discussing in more detail in the following Sect. III, (because if "0" is transmitted, the probability of receiving any photon is exactly 0), the channel is of a "Z" type (Fig. 0.2b).

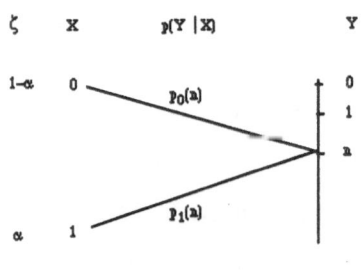

Fig. 0.3a. PPM–Soft decision

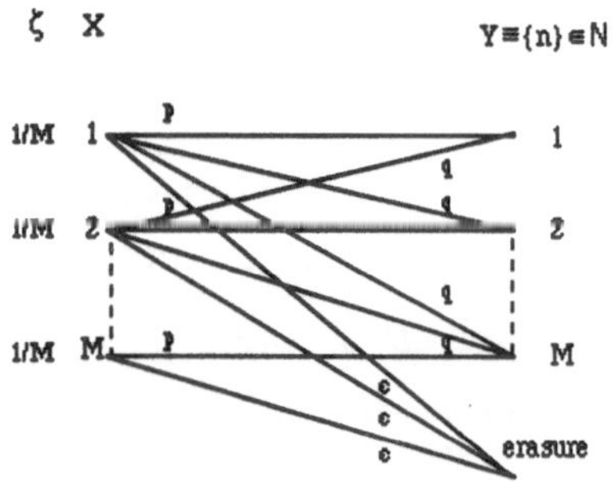

Fig. 0.3b. PPM–Hard decision

In the PPM method, as suggested by Pierce, the signal is a sequence of pulses and the time interval T is divided into M-time slots of duration $\tau = \frac{T}{M}$. In this format, the signal pulse is located in one of these M slots. It can be represented as $\{x_k(t)\}_{k=1}^{M} = A\,F(t - \theta - k\Delta)$, $F(t) = F_0 \text{sinc}(t)$, $\Delta << \theta$.

For example, the input modulation symbol is $\overrightarrow{x} = [x_1, x_2, x_3, ...x_i, ... x_M]$ and the receiver delivers $\overrightarrow{y} = [n_1, n_2, n_3, ...n_i, ...n_M]$.

When the numbers $\{n_i\}_{i=1}^{M} \in \mathbb{N}^M$ the processor is called a *soft decoder* (Fig. 0.3a).

When the decoder acts so that $\overrightarrow{y} = [1,1,1,1,...1,.....1, \varepsilon]$ where ε is the erasure symbol used when no decision was possible, the processing is referred to as M-ary *hard decoding* (Fig. 0.3b). For a noiseless channel $q = 0$.

The final elements at the end of the system are receiver devices. The type of receivers of interest here measure the energy of light via the photoelectric effect. Several types of photodetectors are commonly used in optical communication systems but the photodiode, PIN and avalanche, is the most commonly used because of its superiority over other types of detectors (e.g. photomultiplier) with regard to practical parameters such as quantum efficiency (≥ 0.5), multiplicative gain (≥ 50) and background noise.

The receiver called *direct detection* detects the light intensity and converts the collected photons into an electrical current. Such a detector introduces inevitable noise called *shot noise*: it comes from the statistical process of absorption of photon–electron-hole pair creation. The average intensity of this current is proportional to the average intensity of the light field. Although this system is not sensitive to information which might be contained in the frequency (or phase) of the incident light, it is analyzed in detail in Sect. III because of its practical importance.

When the needed information is contained in the phase (or frequency) of the input signal, the system to use has to be sensitive to the light– amplitude. It belongs to the so called *coherent receiver* class: it consists of the signal light, beam coupler, local oscillator with very high power and frequency tunability, and photodetector. When the frequency of the local oscillator is exactly the same as the incident light, the system is called *homodyne* detection. When the frequency of the local oscillator is different, it is called *heterodyne* detection. An example of the system configuration for coherent detection is shown below in Fig. 0.4.

The heterodyne receiver consists of a photodetector and a laser *local oscillator* which has strong power output and a frequency (ω_L) different from the incoming light (ω_s). The incoming light is mixed with the local laser by a *beam splitter*, and the component of the beat frequency between two light frequencies is detected as an electrical current.

When $\omega_s = \omega_L$, the receiver is then a homodyne one: the incoming light is mixed with the local laser, and the component corresponding to the phase difference between two lights is detected as the result of the interference of two light waves. The output of the photodiode (direct current) is proportional to the phase difference between two light waves.

So if the incoming light is modulated by the phase modulator, the homodyne receiver demodulates the phase modulated signal conveyed by the light wave, because the phase of the local laser is fixed. Such coherent receivers are very important for the application of the quantum communication theory to the practical systems.

A schematic diagram of a so–called *balanced homodyne receiver* is shown in Fig. 0.4. Such a system can remove the noise caused by the quantum noise in the local oscillator light.

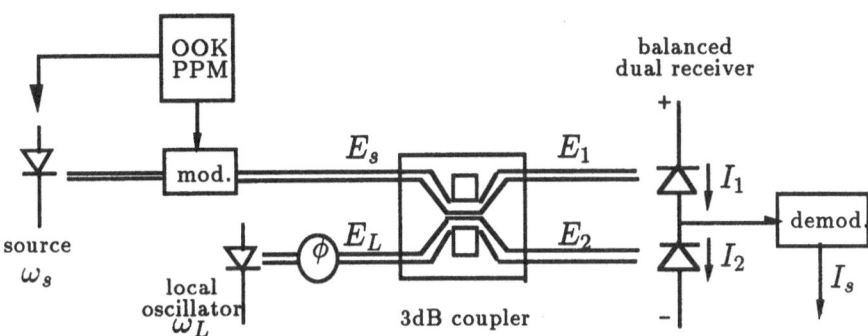

Fig. 0.4. Representation of homodyne and heterodyne communication schemes.

Now, a brief analysis of the heterodyne communication scheme shows that for $\langle E_L \rangle \gg \langle E_s \rangle$, the demodulated intensity is given by $I_s = I_2 - I_1 \sim 2\langle E_L \rangle X_c^{(s)}$ where $X_c^{(s)}$ is one quadrature component of the signal field. Therefore taking into account $\sigma_{X_c^{(s)}} < \delta^2$, the intensity fluctuations $\sigma_{I_s}^2 \sim 4\langle E_L \rangle^2 \sigma_{X_c^{(s)}}$ can be made less than the standard quantum limit. Thus, the homodyne receiver has the highest sensitivity from signal to noise ratio point of view.

Moreover, the heterodyne system has two main features: power sensitivity and high frequency resolution (parameter that characterizes the ability of an apparatus to discriminate different frequency components).

These properties come from the fact that only the part of the beat signal that falls within the receiver bandwidth is selected.

However, there is a 3dB penalty due to the noise generated during the mixing whereas in r.f. reception, the noise is generated before the mixer. This is the another difference between optical and r.f. coherent processing. As will be seen in Sect. II, other criteria than the signal to noise ratio, e.g. the probability of error in detection can characterize the sensitivity limits of the system.

In the following table, are summarized the approximate values of S, average photon number of signal of real amplitude, required to attain probability of error of 10^{-9}.

modulation	receiver	coherent		squeezed [(*)]	
OOK	direct detection	20	(1a)	14	(2a)
	homodyne	36	(1b)	20	(2b)
PPM	direct detection	10		7	
	homodyne	18		10	
PSK	homodyne	9		5	

The table presents some numerical values of the sensitivity limits for some ideal receivers for binary channels using coherent and squeezed input states for various types of signal modulation. (*): calculations are done with the squeezing parameter $r = 0.5$ (see Sect. III.1.3c for comments).

These values are obtained from well–known formula that are recalled below or can be demonstrated from the results given in the following sections. They are obtained from $|\langle q|\phi \rangle|^2$, where $|q\rangle = |n\rangle$ or $|x\rangle, |\beta\rangle$ for direct detection and coherent detections respectively. The states $|\phi\rangle$ are $|\alpha\rangle$ or $|\alpha, \zeta\rangle$ for coherent and squeezed cases respectively.

$(1a): \frac{1}{2}e^{-S}$, $\qquad\qquad (1b): \frac{1}{2}\mathrm{erfc}\left(\sqrt{\frac{S}{2}}\right)$,

$(2a): \frac{1}{2\cosh(r)}e^{-(S-\sinh^2 r)(1+\tan r)}$, $\qquad (2b): \frac{1}{2}\mathrm{erfc}\left(\sqrt{\frac{Se^r}{2}}\right)$,

where $\mathrm{erfc}(x) = 1 - \int_x^\infty \frac{1}{\sqrt{2\pi}}e^{-\frac{t^2}{2}}dt$.

The OOK system uses time signaling pulses which are twice as long as PPM. If the systems are compared at equal signal energy, the PPM system consistently shows better performance. For all types of modulation, use of squeezed light leads to improvement of performance. Here, although the information signals are all classical parameters, an unavoidable noise arises in the detection process such as the heterodyne and homodyne systems. In the semiclassical theory, such noises are interpreted as the shot noise caused by the local laser light. This must be differently interpreted from the quantum mechanical theory point of view. In fact, it is very important to notice that the noise in the detection processes cannot be avoided, even if all devices are the ideal ones.

This section is ended with the mention of a type of receiver which recently stimulated important work: the quantum non–demolition *(QND) receiver.* The principle of the mode of operation is shown below.

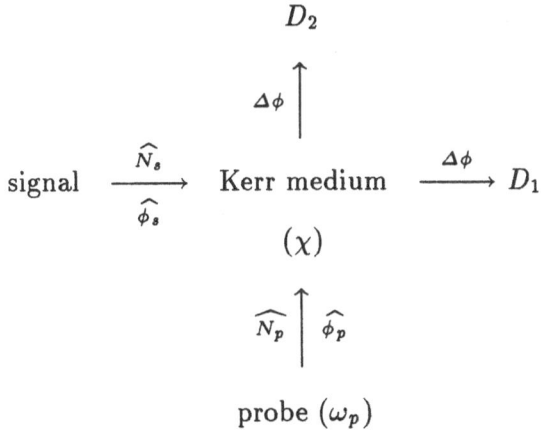

Fig. 0.5. Scheme of a quantum non demolition measurement of the probe field using Kerr effect. D_1 and D_2 are the receivers.

Basically, this type of measurement is characterized by the fact that it makes the observable under study (e.g. position $\hat{x}$ of a particle) independent upon the time. Obviously, its conjugated observable (momentum $\widehat{p}$ of a particle) is being measured with some pertur-bation. Such a property is very important for information trans-mission because it permits repeated measurement on $\widehat{x}$ as its state is not destroyed.

For a lossless Kerr medium, characterized by an $\widehat{H}_I = \hbar\chi\widehat{N}_s\widehat{N}_p$, Hamiltonian of interaction between signal photon number and probe photon number operators $\widehat{N}_s, \widehat{N}_p$ respectively, only the phase of the signal field is changed, the number of photons remains con-stant. Thus, observation of the phase shift of the probe field deter-mines the presence of a photon without destroying the photon.

The accuracy of such a system can be increased by incorporating the Kerr medium in a r.f. cavity and by preparing the probe in a squeezed state.

Because other types of measurement, e.g. direct detection of pho-tons, demolish the observable (number of photons), several practi-cal applications of QND measurement are expected. Among these are the interferometric detection of gravitational waves and also the generation of squeezed states of light.

Summary

Sect. I is a short presentation of the non relativistic quantum–mechanical formalism used in the book. Several types of operators that are of interest in quantum communication are studied. The concepts and results concerning recent progress such as squeezing states or recent ideas about the postulates needed to characterize the measurement of physical quantities, are briefly presented.

In Sect. II, detection and information criteria are analyzed from both points of view, classical as well as quantum mechanical.

Thus, main theorems are recalled and applied to some important practical cases such as the transmission of coherent signals through a channel of Gaussian properties.

In Sect. III, fairly complete results are presented for the case of direct detection which is very important because it is widely used in practice in the detection of signals at low level by a phodetector such as a photomultiplier (or avalanche photodiode).

Finally, in Sect. IV, recent results on the phase operator are presented. Comparison of different models from the point of view of a possible application to optical communications is made.

In conclusion, suggestion for a future work to establish a correspondence between information theory and statistical decision theory in quantum communication, is briefly presented.

I. Elements of the Mathematical Formulation of the Quantum Theory

In this section, we briefly review some elements of the mathematical formalism of quantum mechanics. We then present the main properties of special operators and states that are very useful in quantum optics and communications.

The main object of this section is therefore to fix the notations and to recall the essential results needed in the following. No attempt to be mathematically rigorous has been made. For simplicity, many of the difficult points of the theory are omitted. Only some postulates are used here. The formalism (at the elementary level) used here involves the concepts of vectors, operators, matrices in a space that may have a finite or an infinite number of dimensions.

The interested reader is referred to the bibliography (see for example [1,4,7,13,17]) for further reading.

I.1 States, Operators in Quantum Communication

In the Hilbert space $\mathcal{H}$ we consider here, the number of linearly independent vectors denoted $(|\ \rangle)$ called state–vectors, determines the dimension d of the space where a scalar product $(\langle\ |\ \rangle)$ is defined. Every such $\mathcal{H}$ admits complete orthonormal bases, namely a set of vectors $\{|n\rangle\}_1^d$, $\langle m|n\rangle = \delta_{nm}$ which span $\mathcal{H}$. The fundamental relation is the relation of identity $\mathbb{1} = \sum_{n=1}^d |n\rangle\langle n|$. The observables and time–dependent quantities are represented by linear operators

$\widehat{A}$ that are a transformation of vectors on $\mathcal{H}$. These operators are often non commutative $[\widehat{A}, \widehat{B}] = \widehat{A}\widehat{B} - \widehat{B}\widehat{A} \neq 0$.

In a given representation, the operators can be defined by a matrix of elements $A_{mn} = \langle m|\widehat{A}|n \rangle$ for which the symbol "Tr" denotes the trace operation $\mathrm{Tr}(\widehat{A}) = \sum_n A_{nn}$. The matrix elements obeying the inequality $A_{mn} \leq A_{mm} A_{nn}$. Notice that because of the cyclic property of the trace ($\mathrm{Tr}\,\widehat{A}\widehat{B} = \mathrm{Tr}\,\widehat{B}\widehat{A}$), $\mathrm{Tr}\,[\widehat{A}, \widehat{B}] = 0$ even if $[\widehat{A}, \widehat{B}] \neq 0$. Among the properties required for operators to represent physical quantities, the most important here are the boundness $\|\widehat{A}|\psi\rangle\| \leq C \|\|\psi\rangle\|$ where $C = \sup_{\|\psi\| \leq 1} \|\widehat{A}|\psi\rangle\| < \infty$, the norms of the operator and the state vector being $\|\widehat{A}\| = C$, $\|\|\psi\rangle\| \stackrel{\mathrm{def}}{=} (\langle\psi|\psi\rangle)^{\frac{1}{2}}$ respectively. The self–adjointness $\widehat{A} = \widehat{A}^+$, $\langle\lambda|\widehat{A}|\psi\rangle = (\langle\psi|\widehat{A}|\lambda\rangle)^*$ is the most important property of operators considered here. The unitarity property which reads $\widehat{A}\widehat{A}^\dagger = \widehat{A}^\dagger\widehat{A} = \mathbb{1}$ is sometimes required.

For a self–adjoint, positive ($\langle\psi|\widehat{A}|\psi\rangle \geq 0, \forall|\psi\rangle \in \mathcal{H}$) operator in ($\mathcal{H}, d \to \infty$), having a complete set of eigenvectors and eigenvalues, an important representation of operators is given by the spectral decomposition

$$\widehat{A} = \sum_n \lambda_n |\lambda_n\rangle\langle\lambda_n|$$

with $\lambda_n \in \mathbb{R}, \geq 0$, and such that $\sum_n \lambda_n < \infty$. The integral representation in the definition domain Ω, reads $\widehat{A} = \int_\Omega \lambda dE(\lambda)$, where $dE(\lambda)$ is a spectral family of projection operators.

Let us specify now how the system in $\mathcal{H}$ having a hermitian, time–independent Hamiltonian $\widehat{H}$, behaves dynamically.

The first postulate for the time evolution of the state is given by the Schrödinger equation

$$i\hbar \frac{\partial}{\partial t}|\psi(t)\rangle = \widehat{H}|\psi(t)\rangle$$

($\hbar = 1.042.10^{-34}$ Js), for which the solution is conveniently written in terms of the unitary operator $\widehat{U}(t,t_0) = e^{-\frac{i}{\hbar}(t-t_0)\widehat{H}}$, $|\psi(t)\rangle = \widehat{U}(t,t_0)|\psi(t_0)\rangle$.

Another picture called the *Heisenberg picture* assumes the vectors stationary and treats the operators as time–dependent. These two pictures are simply related so that $|\psi_S(t)\rangle = \widehat{U}(t,t_0)|\psi_H(t)\rangle$ and $\widehat{A}_H(t) = \widehat{U}^\dagger(t,t_0)\widehat{A}_S(t)\widehat{U}(t,t_0)$. Therefore for the conservative system, the time–evolution is given by

$$i\hbar \frac{\partial}{\partial t}\widehat{A}_H(t) = \left[\widehat{A}_H(t), \widehat{H}\right]$$

The third picture we mention here is called *interaction picture*. It is useful when the $\widehat{H}$ may be written as the sum of two terms $\widehat{H} = \widehat{H}_0 + \widehat{H}_I$. In that case, for $\widehat{U}_0(t,t_0) = e^{-\frac{i}{\hbar}(t-t_0)\widehat{H}_0}$, the relations between all these pictures read $\widehat{A}_I(t) = \widehat{U}_0^\dagger(t,t_0)\widehat{A}_S(t)\widehat{U}_0(t,t_0)$ and

$$i\hbar \frac{\partial}{\partial t}|\psi_I(t)\rangle = \widehat{H}_I|\psi_I(t)\rangle$$

In this section, we will begin with summarizing some elementary properties of projection and density operators. We will also establish some basic results concerning the harmonic oscillator.

The main properties of several operators of interest in quantum communication that are used in the subsequent sections, are then recalled. Finally, elements of the quantum measurement theory are briefly presented.

I.1.1 Some Operators

Among the operators that are used here, we will mention the projection, density , probability valued measure operators and some other operators that are essential in the study of harmonic oscillators.

I.1.1a Projection

The class of projection operators is of special importance in quantum communication theory. The properties recalled here will be useful in the following sections

$$\widehat{\Pi}_m^2 = \widehat{\Pi}_m$$
$$\sum_m \widehat{\Pi}_m = \mathbb{1}; \ \widehat{\Pi}_m \widehat{\Pi}_n = 0, \ m \neq n$$
$$\left(\widehat{\Pi}_+\right)^2 = \left(\widehat{\Pi}_m + \widehat{\Pi}_n\right)^2 = \widehat{\Pi}_+ \ \text{iif} \ \widehat{\Pi}_m \widehat{\Pi}_n = \widehat{\Pi}_n \widehat{\Pi}_m = 0$$
$$\left(\widehat{\Pi}_-\right)^2 = \left(\widehat{\Pi}_m - \widehat{\Pi}_n\right)^2 = \widehat{\Pi}_- \ \text{iif} \ \widehat{\Pi}_m \geq \widehat{\Pi}_n$$
$$\left(\widehat{\Pi}_m.\widehat{\Pi}_n\right)^2 = \widehat{\Pi}_m.\widehat{\Pi}_n \ \text{if} \ \left[\widehat{\Pi}_m, \widehat{\Pi}_n\right] = 0$$

I.1.1b Density

An important class of operators that is widely studied in this section, is the class of density operators that can be characterized by the following properties

$$\widehat{\Lambda} = \widehat{\Lambda}^\dagger; \; \text{Tr}\left(\widehat{\Lambda}\right) = 1; \text{Tr}\left(\widehat{\Lambda}^2\right) < 1 \quad (= 1 \text{ for a pure state, } \widehat{\Lambda} = \widehat{\Pi}_{m_0})$$

$$\langle\psi|\widehat{\Lambda}|\psi\rangle \geq 0, \quad \forall|\psi\rangle \subset \mathcal{H}$$

$$\widehat{\Lambda} = \sum_m p_m \widehat{\Pi}_m; \; \widehat{\Pi}_m = |m\rangle\langle m|; \; \widehat{\Pi}_m \widehat{\Pi}_n = 0, m \neq n; \; \sum_m p_m = 1$$

$$\text{Tr}\left(\widehat{\Lambda}\widehat{\Lambda}'\right) = \text{Tr}\left(\widehat{\Lambda}'\widehat{\Lambda}\right)$$

$$\partial\,\text{Tr}\left(f(\widehat{\Lambda})\right) = \text{Tr}\left(\partial\widehat{\Lambda}\partial f(\widehat{\Lambda})\right); \text{Tr}\left(\partial\widehat{\Lambda}\widehat{\Lambda}'\right) = 0,$$

$$\text{for } \forall\partial\widehat{\Lambda} = \partial\widehat{\Lambda}^\dagger \;\to\; \widehat{\Lambda}' = 0$$

$$i\hbar\frac{d\widehat{\Lambda}}{dt} = \left[\widehat{H}, \widehat{\Lambda}\right]; \; \frac{d}{dt}e^{\widehat{A}(t)} \neq \frac{d\widehat{A}(t)}{dt}e^{\widehat{A}(t)}, \; = \text{ if } [\widehat{A}(t), \frac{d\widehat{A}(t)}{dt}] = 0$$

For example, in order that the operator $\widehat{\Lambda} = \xi_0\widehat{\Pi}_0 + \xi_1\widehat{\Pi}_1 = \xi_0|\alpha_0\rangle\langle\alpha_0| + \xi_1|\alpha_1\rangle\langle\alpha_1|$ with $\widehat{\Pi}_i \geq 0, \langle\alpha_0|\alpha_1\rangle = \gamma \in \mathbb{R}$, becomes a density operator $\text{Tr}\,\widehat{\Lambda} = 1 \to \xi_0 + \xi_1 = 1$, and $\widehat{\Lambda} = \lambda_0|\lambda_0\rangle\langle\lambda_0| + \lambda_1|\lambda_1\rangle\langle\lambda_1|$, where $\lambda_0 = \frac{1+\sqrt{1-4p(1-\gamma^2)}}{2}, \lambda_1 = \frac{1-\sqrt{1-4p(1-\gamma^2)}}{2}$, denoting $p = \xi_0\xi_1$ and $|\lambda_0\rangle = (1,1)^t, |\lambda_1\rangle = (1,-1)^t$ for $\xi_0 = \xi_1 = \frac{1}{2}$. Then $\widehat{\Lambda}^{\frac{1}{2}} = \sum_k \sqrt{\lambda_k}|\lambda_k\rangle\langle\lambda_k|$ and $\widehat{\Lambda}^{-1} = \sum_k \frac{1}{\lambda_k}|\lambda_k\rangle\langle\lambda_k|$, where $\lambda_k \neq 0$.

An interesting remark is that the average value of the time derivative of a density operator with a discrete spectrum and which does not explicitly depend on time, is equal to 0.

I.1.1c Positive Operator Valued Measure

An ensemble of M non–negative self–adjoint operators $\widehat{A}_i$ on $\mathcal{H}$ of dimension d satisfying the resolution of identity $\sum_{i=1}^M \widehat{A}_i = \mathbb{1}$ constitutes a positive operator valued measure (p.o.m.). A general pure p.o.m. has all $\widehat{A}_i$ of rank one $\widehat{A}_i = |a_i\rangle\langle a_i|$ but $\langle a_i|a_j\rangle \neq \delta_{ij}$.

Assuming that $\mathrm{Tr}\left(\widehat{A}_i\right)$ is finite, it is then possible to construct an ensemble of density operators such that $\widehat{\Lambda}_i = \dfrac{\widehat{A}_i}{\mathrm{Tr}\left(\widehat{A}_i\right)}$ [37].

I.1.1d Reduced Density Operator

For quantum systems that are independent in some particular measurement processes, it is useful to introduce the concept of the reduced density operator.

Suppose that a system $\mathcal{S}$ is composed of two sub–systems $\mathcal{A}$ and $\mathcal{B}$ so that the Hilbert space $\mathcal{H}_\mathcal{S}$ is a tensor product of two independent spaces $\mathcal{H}_a \otimes \mathcal{H}_b$. Let $\{|a_m\rangle\}$ be a complete orthormal set of vectors spanning $\mathcal{H}_a$ and let $\{|b_n\rangle\}$ be a complete orthormal set of vectors spanning $\mathcal{H}_b$. A ket $|\psi\rangle \equiv |a_m, b_n\rangle = \sum_{m,n} \alpha_{m,n} |a_m\rangle |b_n\rangle$ is a vector of space $\mathcal{H}_\mathcal{S}$, where a density operator $\widehat{\Lambda}$ is defined. An observable $\widehat{A}_a$ of the sub–system $\mathcal{A}$ is described by the matrix elements $\langle a_m|\widehat{A}_a|a_{m'}\rangle = \langle a_m, b_n|\widehat{A}_a \otimes \mathbb{1}_b|a_m, b_n\rangle$ in $\mathcal{H}_a$.

Similarly, another observable $\widehat{B}_b$ of the subsystem $\mathcal{B}$ can also be defined. Notice that $\widehat{A}_a$ and $\widehat{B}_b$ share a common set of eigenstates iif $[\widehat{A}_a, \widehat{B}_b] = 0$. Now, to define the reduced density operator, let us compute the eigenvalue taken by $\widehat{A}_a$ in the total space $\mathcal{H}$

$$
\begin{aligned}
\langle \widehat{A}_a \rangle &= \mathrm{Tr}\left(\widehat{A}_a \widehat{\Lambda}\right) = \mathrm{Tr}\,\widehat{\Lambda}(\widehat{A}_a \otimes \mathbb{1}_b) \\
&= \sum_{m,m',n,n'} \langle a_m, b_n|\widehat{\Lambda}|a_{m'}, b_{n'}\rangle \langle a_{m'}, b_{n'}|\widehat{A}_a \otimes \mathbb{1}_b|a_m, b_n\rangle \\
&= \sum_{m,m',n} \langle a_m, b_n|\widehat{\Lambda}|a_{m'}, b_n\rangle \langle a_{m'}|\widehat{A}_a|a_m\rangle \\
&= \sum_{m,m'} \left(\sum_n \langle a_m, b_n|\widehat{\Lambda}|a_{m'}, b_n\rangle\right) \langle a_{m'}|\widehat{A}_a|a_m\rangle
\end{aligned}
$$

$$= \sum_{m,m'} \langle a_m | \mathrm{Tr}_b \ \hat{\Lambda} | a_{m'} \rangle \langle a_{m'} | \hat{A}_a | a_m \rangle = \sum_{m} \langle a_m | \left(\mathrm{Tr}_b \ \hat{\Lambda} \right) \hat{A}_a | a_m \rangle$$

$$\equiv \mathrm{Tr}_a \ \hat{\Lambda}_a \hat{A}_a$$

where $\langle b_n | b_{n'} \rangle = 1$, iif $n' = n$; $\langle a_m | a_{m'} \rangle = 1$, iif $m' = m$ and the reduced density operator reads $\mathrm{Tr}_b \ \hat{\Lambda} \overset{\mathrm{def}}{=} \hat{\Lambda}_a$.

Another simple application consists in writing the density operator for a beam traveling along the $\vec{z}$ axis, where the $|x\rangle = (1,0)^t$ and $|y\rangle = (0,1)^t$ are the $\vec{x}$–polarized and $\vec{y}$–polarized states respectively. For a state with a polarization vector at some angle ϕ_k with respect to the $\vec{x}$ axis, the density matrix describing a pure state is given by $\hat{\Lambda}_k = \left(\cos \phi_k, \sin \phi_k \right)^t \left(\cos \phi_k, \sin \phi_k \right)$. A mixed state is described by $\hat{\underline{\Lambda}} = \sum_k p_k \hat{\Lambda}_k$.

I.1.1e Operators for a Harmonic Oscillator

The harmonic oscillator in quantum mechanics is described by operators $\hat{x}$ and $\hat{p}$ satisfying the canonical commutation relation (CCR) $[\hat{x}, \hat{p}] = i\hbar$. This (CCR) implies that the space on which these operators act is infinite dimensional and generally these operators should be bounded. It is known that an analogue of the CCR for operators with a discrete bounded spectrum acting on a finite dimensional space, can de proved.

From two operators satisfying $\hat{a}\hat{b} = \epsilon \hat{b}\hat{a}$, and $\hat{a}^d = \hat{b}^d = \mathbb{1}$, where $\epsilon = \exp\left(\frac{2\pi}{d}\right)$, two other operators that are Hermitian, could be formally defined as $\hat{q} = -i\sqrt{\frac{d}{2\pi}} \log \hat{a}$ and similarly $\hat{p} = -i\sqrt{\frac{d}{2\pi}} \log \hat{b}$, for which the matrix elements of the commutator in the basis $\left(\mathcal{B} : \{ |\epsilon^k \rangle \}, \langle \epsilon^k | \epsilon^\ell \rangle = \frac{1}{\sqrt{d}} \exp\left(\frac{2\pi}{d} k\ell\right), k, \ell = 0, ..., d-1 \right)$, read

$$[\hat{q}, \hat{p}]_{k,\ell} = \begin{cases} \frac{d(d-1)}{2}, & \epsilon^{k-\ell} = 1 \\ \frac{d}{\epsilon^{k-\ell}-1}, & \epsilon^{k-\ell} \neq 1 \end{cases}$$

It is noticed that this commutator is such that $\mathrm{Tr}\,[\hat{q}, \hat{p}] = 0$. However, it does reduce to the conventional form in the limit of the continuous spectrum obtained for $d \to \infty$ using the method that replaces the discrete sum by the integral [11].

The Hamiltonian operator for the free particle of mass $m = 1$ is given by

$$\hat{H} = \frac{1}{2}\left(\hat{p}^2 + \omega^2\hat{x}^2\right) \tag{I.1.1a}$$

To find its eigenvalues in a space $d \to \infty$, it is usual to introduce a non–Hermitian operator $\hat{a} = \frac{\omega\hat{x}+i\hat{p}}{\sqrt{2\hbar\omega}}$ and, from the CCR, it is derived that $[\hat{a}, \hat{a}^\dagger] = \mathbb{1}$. Now, with the new definitions

$$\hat{x} = \sqrt{\frac{\hbar}{2\omega}}(\hat{a} + \hat{a}^\dagger)$$

$$\hat{p} = -i\sqrt{\frac{\hbar\omega}{2}}(\hat{a} - \hat{a}^\dagger) \tag{I.1.1b,c}$$

a generalized form of the commutation relation can be obtained. From $\sigma_R^2 \equiv \langle\psi|\hat{R}^\dagger\hat{R}|\psi\rangle \geq 0$, where $\hat{R} = \hat{p} - \zeta\hat{x}$, $\zeta = r\exp(i\phi)$, we easily prove the *Heisenberg inequality*

$$\sigma_p^2\sigma_x^2 - (\Re e\langle\psi|\hat{x}\hat{p}|\psi\rangle)^2 \geq \frac{\hbar^2}{4} \tag{I.1.1d}$$

where

$$\sigma_{\hat{O}}^2 \stackrel{\text{def}}{=} \langle\psi|\hat{O}^2|\psi\rangle - \langle\psi|\hat{O}|\psi\rangle^2 = \langle\hat{O}^2\rangle - \langle\hat{O}\rangle^2 \tag{I.1.1e}$$

For any function $F(\hat{p})$ that can be expanded in Taylor series,

$$\sigma_F^2 \sigma_x^2 - (\Re e\langle\psi|\hat{x}F(\hat{p})|\psi\rangle)^2 \geq \frac{\hbar^2}{4}\langle|\frac{dF}{d\hat{p}}|^2\rangle \qquad (\text{I.1.1}f)$$

Another uncertainty relation connecting energy and time can be written in the form $\sigma_E^2 \sigma_t^2 \geq \hbar$. It is however is of different interpretation because t is a parameter in the sense that for a system evolving under a Hamiltonian $\hat{H}$, the state evolves so that $e^{-i\hat{H}t'}|e(t)\rangle = |e(t + t')\rangle$. We cannot associate any operator with such parameter.

An entropic formulation of the uncertainty relation will be established at the end of this section.

Now if $|n\rangle$ is an eigenvector of $\hat{N} = \hat{a}^\dagger\hat{a}$, it is also an eigenvector of

$$\hat{H}|n\rangle = (\hat{a}^\dagger\hat{a} + \frac{1}{2})\hbar\omega|n\rangle = \left(\hat{N} + \frac{1}{2}\right)\hbar\omega|n\rangle = (n + \frac{1}{2})\hbar\omega|n\rangle$$

By a suitable choice of arbitrary constant, we find

$$\begin{cases} \hat{a}^k|n\rangle = \sqrt{\frac{n!}{(n-k)!}}|n - k\rangle \\ \hat{a}^{\dagger k}|n\rangle = \sqrt{\frac{(n+k)!}{n!}}|n + k\rangle \end{cases} \qquad (\text{I.1.2})$$

The basis $\{|n\rangle\}$ of the Fock space obeys the resolution of identity $\sum_n |n\rangle\langle n| = \mathbb{1}$. It is also useful to notice that $[\hat{a}^2, \hat{a}^{\dagger 2}] = 4\hat{a}^\dagger\hat{a} + 2$.

Remarks:

- the state $|\varphi\rangle = \widehat{a}^{\dagger}|0\rangle$ cannot be *normalized*

- the operator $\widehat{a}$ cannot be written in the form $\widehat{a} = \widehat{U} f(\widehat{N})$ such that $\widehat{U}$ is unitary and $f(\widehat{N}) = \sqrt{\widehat{N}+1}$. This is because the spectrum of $\widehat{N}$ does not include the negative integers. An "approximate" polar decomposition $\widehat{U} = \widehat{C} + i\widehat{S}$, will be discussed below in Sect. IV, when dealing with the quantum description of the optical phase operator.

- the operators $\widehat{a}$ and $\widehat{a}^{\dagger}$ cannot be *inverted*. Generalized inverses are however introduced in a recent work [35].

Moreover, a very important formula, demonstrated by Glauber [3], that will be often used in this section is: for $\widehat{A}, \widehat{B}$ verifying $\left[\widehat{A}, [\widehat{A}, \widehat{B}]\right] = \left[\widehat{B}, [\widehat{A}, \widehat{B}]\right] = 0$, we have

$$\exp\left(\widehat{A} + \widehat{B}\right) = \exp\left(\widehat{A}\right)\exp\left(\widehat{B}\right)\exp\left(-\tfrac{1}{2}[\widehat{A}, \widehat{B}]\right) \qquad (\text{I}.1.3)$$

The proof is based on the solution of the equation $\frac{df(\xi)}{d\xi} = \left(\widehat{A} + e^{\xi\widehat{A}}\widehat{B}e^{-\xi\widehat{A}}\right)f(\xi) = \left((\widehat{A} + \widehat{B}) + \xi[\widehat{A}, \widehat{B}]\right)f(\xi)$ where $f(\xi) = e^{\xi\widehat{A}}e^{\xi\widehat{B}}$ and $f(0) = 1$. The solution for $\xi = 1$ takes the form (I.1.3).

In the case that the time–dependent part of the Hamiltonian a harmonic oscillator can be seen as a perturbation of the total Hamiltonian $\widehat{H} = \widehat{H}_0 + \beta\widehat{h}(t)$, $(\widehat{H}_0|n(0)\rangle = (n + \tfrac{1}{2})|n(0)\rangle$, the approximation of the solution of the Schrödinger equation for the time–dependent number state, can be derived from the expansion in powers of β

$$
\begin{cases}
|n(t)\rangle = \sum_k \sum_\ell \beta^\ell b_k^{(\ell)}(t) \exp\left(-\frac{n+\frac{1}{2}}{\hbar} t\right) |n(0)\rangle \\[2mm]
i\hbar \frac{d}{dt} b_k^{(\ell)}(t) = \sum_j e^{i\omega_{jk} t} \widehat{h}_{jk}(t) b_k^{((\ell-1))}(t), \quad b_\ell^{(0)}(t) = \delta_{\ell n} \\[2mm]
\omega_{jk} = \frac{E_j - E_k}{\hbar} \\[2mm]
\widehat{h}_{jk}(t) = \langle j|\widehat{h}(t)|k\rangle
\end{cases}
\tag{I.1.4}
$$

To treat a more general form than (I.1.1a) including time–varying frequency, let the Hamiltonian be given by

$$
\widehat{H}(t) = \frac{1}{2}\widehat{p}^2 + \frac{1}{2}\omega^2(t)\widehat{x}^2
\tag{I.1.5}
$$

then it can be shown that by controlling $\omega(t)$, it is possible to obtain squeezing properties for the ground state.

The method to prove this result is based on the properties of the time–dependent Hermitian invariant operator $\widehat{I}(t)$ the solution of the evolution equation $\frac{\partial}{\partial t}\widehat{I}(t) = -i[\widehat{H}(t), \widehat{I}(t)]$, $(\hbar = 1)$ and proved to be of a quadratic form

$$
\widehat{I}(t) = \frac{1}{2}\left[\frac{1}{\rho(t)^2}\widehat{x}^2 + \left(\rho(t)\widehat{p} - \frac{d}{dt}\rho(t)\widehat{x}\right)^2\right]
\tag{I.1.6}
$$

where the scalar $\rho(t)$ is the solution of the differential equation

$$
\frac{d^2}{dt^2}\rho(t) + \omega^2(t)\rho(t) - \frac{1}{\rho^3(t)} = 0
\tag{I.1.7}
$$

with the conventional initial conditions.

Rewriting the invariant operator in the form $\widehat{I}(t) = \widehat{A}^\dagger(t)\widehat{A}(t) + \frac{1}{2}$ will be useful to define the time–dependent creation and annihilation operators obeying the commutator $[\widehat{A}(t), \widehat{A}^\dagger(t)] = \mathbb{1}$, where

$$\hat{A}(t) = \frac{i}{\sqrt{2}} \left[\rho(t)\hat{p} + \left(\frac{1}{i\rho(t)} - \frac{d}{dt}\rho(t) \right)\hat{x} \right] \qquad (I.1.8)$$

Now, it is possible to expand the eigenstates $|n(t)\rangle$ of the Schrödinger equation $i\frac{\partial}{\partial t}|n(t)\rangle = \hat{H}(t)|n(t)\rangle$ in terms of the orthonormal eigenstates $|i(t)\rangle$ of $\hat{I}(t)$ of discrete time–independent spectrum $\gamma_n = n + \frac{1}{2}$. For the system in the ground state $|0(t)\rangle$, it is found that $\sigma_x^2(t) = \frac{1}{2}\rho^2(t)$, $\sigma_p^2(t) = \frac{1}{2}\left(\frac{1}{\rho^2(t)} + \left(\frac{d}{dt}\rho(t) \right)^2 \right)$ [36], where σ_x^2 and σ_p^2 are the standard deviations of the operators $\hat{x}$ and $\hat{p}$ derived as usual from $\hat{A}(t)$ and $\hat{A}^\dagger(t)$ (see (I.1.1b,c)).

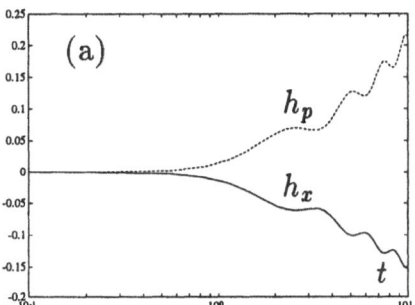 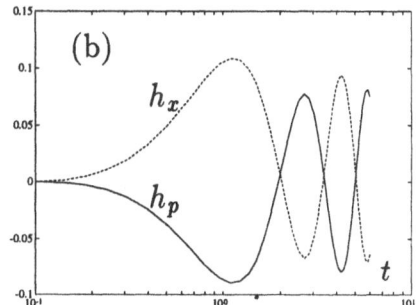

Fig. I.1. (a). Time evolution of the reduced variances $h_i = \sigma_i^2 - \frac{1}{2}$, ($i = x, p$) for weak time–dependent harmonic oscillator. The initial conditions are $\frac{d}{dt}\rho(t)|_{t=0} = 0$, $\rho(0) = 1$. $\omega^2(t) = 1 + \lambda t$. (b). Time evolution of the reduced variances $h_i = \sigma_i^2 - \frac{1}{2}$, ($i = x, p$) for weak time–dependent harmonic oscillator. The initial conditions are $\frac{d}{dt}\rho(t)|_{t=0} = 0$, $\rho(0) = 1$. $\omega^2(t) = 1 + \frac{\lambda}{t}$ ($\lambda = 0.01$).

It is clearly seen in Figs. I.1, that a squeezing can be realized on one quadrature component by an appropriate choice of $\omega(t)$. Other initial conditions will lead to different curves but the squeezing behavior will basically be the same.

Another method based on the solution of the Schrödinger equation $i\frac{\partial \psi(x,t)}{\partial t} = \hat{H}\psi(x,t)$, where the Hamiltonian is written in the form $\hat{H} = -\frac{1}{2}\frac{\partial^2}{\partial x^2} + \frac{\omega(t)^2}{2}\hat{x}^2$, $(\hbar = 1, m = 1)$ can also be utilized. The solution $\psi(x,t) = \int G(x,t;x',t)\psi(x',t)dx'$ is expressed in terms of the Green function

$$G(x,t;x',t) = \exp\left(\frac{i}{2}\left[a(t)x^2 + 2b(t)xx' + c(t)\right]\right)$$

where $a(t) = \frac{1}{Z(t)}\frac{dZ(t)}{dt}$, $b(t) = \frac{1}{Z(t)}$ and $c(t) = i\log Z(t) - \int_0^t b(\theta)^2 d\theta$ and such that $(\lim_{t\to t'} G(x,t;x',t') = \delta(x-x'))$ [2].

The function $Z(t)$ being solution of the ordinary differential equation $\frac{d^2 Z(t)}{dt^2} + \omega(t)^2 Z(t) = 0$ taking into account the boundary conditions $Z(t=t') = 0$, $\frac{dZ(t)}{dt}|_{t=t'} = 1$.

I.2 Representations

The choice of the observable used to represent state vectors and operators is not unique. For example, for a particle moving in one dimension, the momentum or the coordinate would serve as a suitable observable in representing state vectors and operators in matrix form. For a type of calculations, one representation may be more convenient than the other.

In this section, we will consider the most frequently used representations.

I.2.1 Coherent States

The coherent states are eigenvectors of the operator $\hat{a}$. For any $\alpha \in \mathbb{C}$, let us define the state–vector

$$|\alpha\rangle = \sum_{n=0}^{\infty} \frac{\alpha^n}{\sqrt{n!}} \exp\left(-\frac{|\alpha|^2}{2}\right) |n\rangle = \hat{D}(\alpha)|0\rangle \qquad (I.2.1a)$$

for which

$$\hat{a}|\alpha\rangle = \alpha|\alpha\rangle$$
$$\langle\alpha|\alpha^* = \langle\alpha|\hat{a}^\dagger \qquad (I.2.1b)$$

The normalization is simply derived from (I.2.1a)

$$\frac{1}{\pi} \int |\alpha\rangle\langle\alpha| d(\Re e\,\alpha) d(\Im m\,\alpha)$$

$$= \frac{1}{\pi} \sum_{n,m} \frac{1}{\sqrt{n!m!}} \int \exp(-|\alpha|^2)\alpha^m \alpha^{*n} |m\rangle\langle n| d(\Re e\,\alpha) d(\Im m\,\alpha)$$

$$= \sum_m \frac{1}{m!} \left[\int 2|\alpha| d(|\alpha|)|\alpha|^{2m} \exp\left(-|\alpha|^2\right) \right] |m\rangle\langle m| = \sum_m |m\rangle\langle m| = \mathbb{1}$$

which yield the probability that the energy is E_n of a Poisson form.

The scalar product of such states

$$\langle\alpha|\beta\rangle = \exp\left(\alpha^*\beta - \frac{|\alpha|^2}{2} - \frac{|\beta|^2}{2}\right) \qquad (I.2.1c)$$

The following equalities can also be readily derived

$$\begin{cases} \langle\alpha|\hat{x}|\alpha\rangle = \sqrt{\frac{2\hbar}{\omega}}\, \Re e\,\alpha \\ \langle\alpha|\hat{p}|\alpha\rangle = \sqrt{2\hbar\omega}\, \Im m\,\alpha \\ \Re e\langle\psi|\hat{x}\hat{p}|\psi\rangle = 0 \\ \sigma_x^2 = \sigma_p^2 = \frac{\hbar}{2} \end{cases} \qquad (I.2.2)$$

Another type of uncertainty that we here consider is the so–called *entropic uncertainty relation* which is given by a lower bound of the entropies $S_P(\psi)$, $S_Q(\psi)$ of the two operators $\hat{P}, \hat{Q}$ in the form [28]

$$\inf_{|\psi\rangle} (S_P(\psi) + S_Q(\psi)) \geq -2\log \left(\max_{j,k} |\langle p_j|q_k\rangle| \right) \qquad (I.2.3)$$

where $\{|p_j\rangle\}$, $\{|q_k\rangle\}$ are the corresponding complete sets of normalized eigenvectors of $\hat{P}, \hat{Q}$ respectively, in a Hilbert space $\mathcal{H}$ [23]. For a one dimensional particle $\hat{X}, \hat{P}_x \equiv \hat{P}$, defined so that $\hat{X}|x\rangle = x|x\rangle$, $\hat{P}|p\rangle = p|p\rangle$, we have $\langle x|p\rangle = \frac{1}{\sqrt{h}}\exp(\frac{ipx}{\hbar})$. Suppose $|\psi\rangle$ be a given pure state. Setting $\psi_1(x) = \langle x|\psi\rangle$, it is easy to obtain

$$\begin{cases} \psi_1(x) = \displaystyle\int \langle x|p\rangle\langle p|\psi\rangle dp = \frac{1}{\sqrt{h}}\int \exp(\frac{ipx}{\hbar})\psi_2(p)dp \\ \psi_2(p) = \displaystyle\int \langle p|x\rangle\langle x|\psi\rangle dx = \frac{1}{\sqrt{h}}\int \exp(\frac{-ipx}{\hbar})\psi_1(x)dx \end{cases}$$

where $\int |x\rangle\langle x|dx = 1$, $\int |p\rangle\langle p|dp = 1$, $\psi_2(p) = \langle p|\psi\rangle$, and $\langle x|\hat{P}|\psi\rangle = -i\hbar\frac{\partial}{\partial x}\langle x|\psi\rangle$. For $|\psi_1(x)|^2 = \mathcal{N}(\mu_x, \sigma_x)$, we obtain $|\psi_2(p)|^2 = \mathcal{N}(\mu_p, \sigma_p)$. The entropic formulation will therefore be given by

$$\begin{aligned} U(\hat{X}, \hat{P}; \psi) &= S_\psi(\hat{X}) + S_\psi(\hat{P}) \\ &= \frac{1}{2}\log(2\pi e\sigma_x^2) + \frac{1}{2}\log(2\pi e\sigma_p^2) \\ &= \log(2\pi e) + \log(\sigma_x\sigma_p) \\ &\geq 1 + \log\pi \end{aligned} \qquad (I.2.4)$$

where we used the well–known result that the differential entropy of a Gaussian distribution $\mathcal{N}(\mu, \sigma^2)$ is $\frac{1}{2}\log(2\pi e\sigma^2)$.

For a coherent state $|\psi\rangle = |\alpha\rangle$, where we set $\hbar = 1$ and $\sigma_x = \sigma_p = \frac{1}{\sqrt{2}}$, we obtain $U = U_m = 1 + \log \pi$.

Remark: This formulation seems more appropriate than that given by (I.1.1d), and indeed more general as it includes the conventional inequality for the Gaussian case.

I.2.2 Unitary Displacement Operator

The main properties of $\widehat{D}(\alpha)$, the displacement operator that are often used for the calculations

$$\widehat{D}(\alpha) = \exp(\alpha\widehat{a} - \alpha^*\widehat{a}^\dagger) = e^{-\frac{1}{2}|\alpha|^2} \exp(\alpha\widehat{a}^\dagger)\exp(-\alpha^*\widehat{a})$$
$$\widehat{D}(\alpha)|\gamma\rangle = \exp(i\Im(\alpha\gamma^*))|\alpha + \gamma\rangle$$
$$\widehat{D}(\alpha)\widehat{D}(\beta) = \widehat{D}(\alpha + \beta)\exp(\alpha\beta^* - \alpha^*\beta)$$
$$\mathrm{Tr}\left[\widehat{D}(\alpha)\widehat{D}(\beta)\right] = \pi\delta(\alpha - \beta)$$
$$\widehat{D}(\alpha)\widehat{a}\widehat{D}^\dagger(\alpha) = \widehat{a} - \alpha$$
$$\widehat{D}(\alpha)\widehat{a}^\dagger\widehat{D}^\dagger(\alpha) = \widehat{a}^\dagger - \alpha^*$$

$$(I.2.5)$$

The s–ordered operator is $\widehat{D}(\alpha, s) = \exp(\frac{s}{2}|\alpha|^2)\exp(\alpha\widehat{a}^\dagger - \alpha^*\widehat{a})$. To obtain the normal ordering form, we must set $s = 1$.

I.2.3 Squeezed States

Squeezed states constitute another class of states [27]. They are in some sense the generalization of the coherent states [21] and are conveniently generated by using the unitary squeeze operator [16]

$$\widehat{S}(\zeta) = e^{\frac{\zeta^*\widehat{a}^2 - \zeta\widehat{a}^{\dagger 2}}{2}} = e^{-\frac{|\zeta|^2}{4}}e^{\frac{\zeta^*}{2}\widehat{a}^2 - \frac{\zeta}{2}\widehat{a}^{\dagger 2} + \frac{|\zeta|^2}{2}\widehat{a}^\dagger\widehat{a}} \qquad (I.2.6)$$

where $\zeta = r\exp(i\phi)$ is an arbitrary complex parameter.

A useful definition of the squeezed states as a transformation of coherent states is

$$|\alpha, \zeta\rangle = \hat{S}(\zeta)|\alpha\rangle = \hat{S}(\zeta)\hat{D}(\alpha)|0\rangle \tag{I.2.7}$$

With this notation, a coherent state (not a vacuum) is squeezed.

Another construction which first squeezes a vacuum state and then displaces the vacuum squeezed state called *ideal squeezing*, can be defined where the identity relation

$$\frac{1}{\pi}\int |\alpha, \zeta\rangle\langle\alpha, \zeta| d(\Re e\,\alpha)d(\Im m\,\alpha) = \mathbb{1}$$

since the squeeze operator is unitary. It can be seen that these states correspond to the equality for the Heisenberg relation with $\Re e\langle\psi|\widehat{px}|\psi\rangle = \frac{1}{\tan(\phi)}$.

These states can also be defined from the number states $|n, \zeta\rangle = \hat{S}(\zeta)|n\rangle$ yielding the density matrix $\hat{\rho} = \sum_{n=0}^{\infty} p_n |n, \zeta\rangle\langle n, \zeta| = \int \mathcal{P}_t(\alpha)|\alpha, \zeta\rangle\langle\alpha, \zeta| d^2\alpha$ with $p_n = \frac{N^n}{(1+N)^{1+n}}$ and the thermal quasi-probability distribution $\mathcal{P}_t(\alpha) = \frac{1}{\pi N}\exp(-\frac{|\alpha|^2}{N})$. To simplify the calculations, it will often be assumed that $\zeta \equiv r \in \mathbb{R}$. The connecting relations with Yuen's notations for a TCS $|\beta; \mu, \nu\rangle$ [12] are $\mu = \cosh(r)$, $\nu = \exp(i\phi)\sinh(r)$ and $\hat{b} = \mu\hat{a} + \nu\hat{a}^\dagger$. The scalar product of two squeezed states is given by

$$\langle\alpha_1, r_1|\alpha_2, r_2\rangle = \frac{1}{\sqrt{\cosh(r_2 - r_1)}} \times$$

$$\exp\left(-\frac{|\alpha_1|^2 + |\alpha_2|^2}{2} - \frac{[\alpha_1^{*2}e^{-i\phi} - \alpha_2^2 e^{-i\phi}]\tanh(r_2 - r_1)}{2} + \frac{\alpha_1^*\alpha_2}{\cosh(r_2 - r_1)}\right) \tag{I.2.8}$$

We may also define [32]

$$|\alpha, 0\rangle = |\alpha\rangle$$
$$\widehat{A}_\zeta \overset{\text{def}}{=} \widehat{S}(\zeta)\widehat{a}\widehat{S}^\dagger(\zeta) = \widehat{a}\cosh(r) + \widehat{a}^\dagger \exp(i\phi)\sinh(r)$$
$$\widehat{A}_\zeta^\dagger \overset{\text{def}}{=} \widehat{S}(\zeta)\widehat{a}^\dagger \widehat{S}^\dagger(\zeta) = \widehat{a}^\dagger \cosh(r) + \widehat{a}\exp(-i\phi)\sinh(r) \qquad \text{(I.2.9a, b, c, d)}$$
$$\widehat{S}^\dagger(\zeta)f\left(\widehat{a}^\dagger, \widehat{a}\right)\widehat{S}(\zeta) = f\left(\widehat{A}_\zeta^\dagger, \widehat{A}_\zeta\right)$$

for any function that can be expanded in power of $\widehat{a}^\dagger$ and $\widehat{a}$. There-fore, we can set the annihilation and displacement operators $\widehat{A}_\zeta$ and $\widehat{D}_\zeta(\alpha)$ in the "coherent" states form similar to (I.2.1b) and (I.2.5)

$$\widehat{A}_\zeta |\alpha, \zeta\rangle = \alpha |\alpha, \zeta\rangle$$
$$\widehat{D}_\zeta(\alpha) = \exp\left(\alpha \widehat{A}_\zeta^\dagger - \alpha^* \widehat{A}_\zeta\right) \qquad \text{(I.2.9e, f)}$$

To extend this formulation to the case of squeezed states with thermal noise, we follow the presentation [24,30] and define the operator

$$\widehat{T}(\theta) = \exp\left(-\theta\left(\widehat{a}\widehat{b} - \widehat{a}^\dagger \widehat{b}^\dagger\right)\right) \qquad \text{(I.2.10a)}$$

where $\sinh^2 \theta = N_t$ is the number of thermal photons. The "aux-iliary" states and operators are introduced to simplify the calcula-tions. The notations are

$$\widehat{b}|\widetilde{\beta}\rangle = \widehat{b}\widehat{D}(\beta)|\widetilde{0}\rangle$$

The so-called thermofield representation is useful: the initial Hilbert space $\mathcal{H} : \widehat{a}, |n\rangle, \dots$ is enlarged with an auxiliary Hilbert space $\mathcal{K} : \widehat{b}, |\widetilde{m}\rangle, \dots$ to have the basis $|n, \widetilde{m}; \theta\rangle$ and the following

$$|0(\theta)\rangle = \widehat{T}(\theta)|0, \widetilde{0}\rangle, \quad |\alpha(\theta)\rangle = \widehat{D}(\alpha)|0(\theta)\rangle$$

$$\hat{a}(\theta) = \hat{a}\cosh(\theta) + \hat{b}^\dagger \sinh(\theta), \ \hat{b}(\theta) = \hat{b}\cosh(\theta) + \hat{a}^\dagger \sinh(\theta)$$

$$\hat{a}^\dagger \hat{a} \ \hat{b}^\dagger \hat{b} |n, \tilde{m}; \theta\rangle = N_c N_t |n, \tilde{m}; \theta\rangle$$

As for the usual squeezed states, we may define squeezed thermal states

$$|\alpha, \zeta, \theta\rangle_t = \hat{D}(\alpha)\hat{S}(\zeta)\hat{T}(\theta)|0\rangle \otimes |\tilde{0}\rangle \tag{I.2.10b}$$

whose properties differ from those of thermalized squeezed states defined as

$$|\alpha, \theta, \zeta\rangle_T = \hat{D}(\alpha)\hat{T}(\theta)\hat{S}(\zeta)|0\rangle \otimes |\tilde{0}\rangle \tag{I.2.10c}$$

In particular, calculations of variances of the two components of the field (X_c and X_s), yield

$$\begin{cases} \Delta_1 = \sigma_{1t}^2 - \sigma_{1T}^2 = \frac{N}{2}\left(1 - \frac{\exp(-2r)}{2}\right) \\ \Delta_2 = \sigma_{2t}^2 - \sigma_{2T}^2 = \frac{N}{2}\left(1 - \frac{\exp(2r)}{2}\right) \end{cases} \tag{I.2.10d}$$

I.2.4 Number

Like the coherent states representation, the number representation is very useful and widely utilized for interpretation of photoncounting observations. From

$$\hat{N}|n\rangle = N|n\rangle, \ \langle n|m\rangle = \delta_{nm}, \ \sum_{n=0}^{\infty} |n\rangle\langle n| - \mathbb{1} \tag{I.2.11}$$

$$\langle n|\alpha\rangle = \frac{\alpha^n}{\sqrt{n!}} e^{-\frac{|\alpha|^2}{2}}$$

we easily prove $\langle n|\hat{p}|n\rangle = \langle n|\hat{x}|n\rangle = 0$. Then, making use of the identity relation of the states $\{|n\rangle\}$, we obtain for $m \leq n$

$$\langle m|\widehat{D}(\alpha)|n\rangle = \exp\left(-\frac{|\alpha|^2}{2}\right)\sum_{\ell}\langle m|\exp(\alpha\widehat{a}^\dagger)|\ell\rangle\langle\ell|\exp(-\alpha^*\widehat{a})|n\rangle$$

$$= \sqrt{\frac{m!}{n!}}\,(-\alpha^*)^{(n-m)}\exp\left(-\frac{|\alpha|^2}{2}\right)L_m^{(n-m)}\left(|\alpha|^2\right)$$

$$(I.2.12)$$

For example, a thermalized number state will be conveniently defined as $|n,\theta\rangle = \widehat{T}(\theta)|n\rangle$.

Similarly, the squeezed number state being written in the usual form $|m,\zeta\rangle = \widehat{S}(\zeta)|m\rangle$, the calculation of the scalar product yield

$$\langle n|\widehat{S}(\zeta)|m\rangle = \begin{cases} Q_1 + Q_2 & m \le n, \quad m,n \text{ even} \\ Q_2 & m > n \\ Q_1 & m \le n, \quad m,n \text{ odd} \end{cases} \qquad (I.2.13)$$

where for $\zeta \in \mathbb{R}$ [12]

$$\begin{cases} Q_1 = \sqrt{\dfrac{(2\sinh(r))^{n-m}n!}{\frac{n-m}{2}(\cosh(r))^{m+n+1}m!}} \\ Q_2 = (-1)^n\sqrt{\dfrac{n!m!(\tanh(r))^{m+n}}{\cosh(r)2^{n+m}}} \end{cases}$$

The number representation of the squeezed state is also easily derived

$$\langle n|\alpha,\zeta\rangle = \sum_m\langle n|\widehat{S}(\zeta)|m\rangle\langle m|\widehat{D}(\alpha)|0\rangle = \sqrt{\frac{\left[\frac{e^{i\phi}\tanh(r)}{2}\right]^n}{n!\cosh(r)}}$$

$$(I.2.14)$$

$$\times\, e^{-\frac{|\alpha|^2+\alpha^*e^{i\phi}\tanh(r)}{2}}\,H_n\left(\frac{\alpha+\alpha^*e^{i\phi}\tanh(r)}{\sqrt{2e^{i\phi}\tanh(r)}}\right)$$

I.2.5 Coordinate

This representation is very useful for the interpretation of the homodyne measurements [10].

From: $\hat{x}|x\rangle = x|x\rangle$, $\int_{-\infty}^{\infty} |x\rangle\langle x|dx = \mathbb{1}$, $\hbar = 1$, it is proved the following equalities

$$\langle x|\alpha\rangle = \left(\frac{\omega}{\pi}\right)^{\frac{1}{4}} \exp\left(-\frac{\omega x^2}{2} + \alpha\sqrt{2\omega}x - \alpha\,\Re e\,\alpha\right) \qquad (\text{I.2.15})$$

$$\langle n|x\rangle = \frac{1}{\pi}\int \langle n|\alpha\rangle\langle\alpha|x\rangle d(\Re e\,\alpha)d(\Im m\,\alpha)$$

$$= (\omega)^{\frac{1}{4}}\frac{1}{\sqrt{2^n n!\sqrt{\pi}}}\exp\left(-\frac{\omega x^2}{2}\right) H_n\left(x\sqrt{\omega}\right) \qquad (\text{I.2.16})$$

$$\langle x|p\rangle = \frac{1}{\sqrt{2\pi}}\exp\left(ipx\right) \qquad (\text{I.2.17})$$

In the coordinate representation, the operator $\mathbb{1} \times p_x$ is the partial differential operator $-i\dfrac{\partial}{\partial x}$. Similarly, the operator $\dfrac{1}{p_x}$ is the integral $i\displaystyle\int_{-\infty}^{x} d\xi$. Notice that the defined integral operator $\hat{P}\psi(x) = \int_0^x \psi(u)du$ for $x \in [0,1]$ has a norm $\|\hat{P}\| = \frac{2}{\pi}$. On the other hand, for $\zeta \in \mathbb{R}$

$$\langle x|\alpha, r\rangle = \left(\frac{2\exp(2r)}{\pi}\right)^{\frac{1}{4}} \exp\left(-x^2 + 2\alpha\exp(r)x - \frac{|\alpha|^2 + \alpha^2}{2}\right)$$

$$(\text{I.2.18})$$

For $\zeta \in \mathbb{C}$, these expressions take of a more complicated form. It can however be shown [29] that the Gaussian shapes of (I.2.15,18) hold.

I.2.6 Momentum

As was done previously, we set $\hat{p}|p\rangle = p|p\rangle$, $\int_{-\infty}^{\infty} |p\rangle\langle p| dp = \mathbb{1}$. We easily obtain

$$\langle\alpha|p\rangle = \left(\frac{1}{\pi\omega}\right)^{\frac{1}{4}} \exp\left(-\frac{p^2}{2\omega} - \frac{2i\alpha p}{\sqrt{2\omega}} + i\alpha\,\Im m\,\alpha\right) \qquad (\text{I.2.19})$$

$$\langle n|p\rangle = \int \langle n|x\rangle\langle x|p\rangle dx = \frac{1}{\sqrt{2^n n!\sqrt{\pi\omega}}} \exp\left(-\frac{p^2}{2\omega}\right) H_n\left(\frac{p}{\sqrt{\omega\hbar}}\right)$$
$$(\text{I.2.20})$$

$$\langle p|\alpha, r\rangle = \frac{1}{\sqrt{\sqrt{\pi}\cosh(r)}} \exp\left(-\frac{|\alpha|^2 + p^2}{2} - \frac{i\alpha^2\tanh(r)}{2}\right)$$

$$\sum_{n=0}^{\infty} \frac{\sqrt{\left[\frac{i\tanh(r)}{2}\right]^n}}{n!} H_n(p) H_n\left(\frac{\alpha}{\sqrt{i\sinh(2r)}}\right)$$
$$(\text{I.2.21})$$

In the momentum representation, the operator $\mathbb{1} \times x$ is the partial differential operator $i\dfrac{\partial}{\partial p}$. Similarly, the operator $\dfrac{1}{x}$ is the integral operator $-i\displaystyle\int_{-\infty}^{p} d\pi$.

I.2.7 Phase

The problems involving the hermiticity of the quantum phase operator have been of interest since the early study of the quantum theory [6,8,22]. A new interest has recently been stimulated by the development of the squeezed states of light.

As a brief introduction, let us define in analogy with the coherent states (I.2.1a)

$$\begin{cases} \hat{\Phi}|\phi\rangle = \phi|\phi\rangle \\ \\ |\phi\rangle = \sqrt{1 - |z|^2} \displaystyle\sum_{n=0}^{\infty} z^n |n\rangle \end{cases}$$

the phase state being expanded in number states representation in order to have $\langle \phi | \phi \rangle = 1$, where $z = |z| \exp(i\phi)$, $|z| < 1$. It is clearly seen, from the following relations, that some problems, in connection with the normalization of the phase states, arise when $|z| \to 1$, $0 \le \phi \le 2\pi$

$$\begin{cases} \langle \phi | \hat{a}^\dagger \hat{a} | \phi \rangle = \langle \hat{N} \rangle \to \infty \\ \langle \phi | \left(\hat{a}^\dagger \hat{a} \right)^2 | \phi \rangle \to \infty \\ |\langle \psi | \phi \rangle|^2 \to \delta(\theta - \phi) \\ |\langle \phi | n \rangle|^2 \to 0 \end{cases}$$

where we assumed that $|z_i| = |z|$.

Notice the correspondence $|z| \to 1 \iff \langle \hat{N} \rangle \to \infty$.

In fact, for $|\Psi\rangle = \sum_{m=0}^{\infty} e^{im\theta} |m\rangle$ which is the usual phase state, the probability density for obtaining a phase value $\varphi = \theta - \phi$ will be given by [see ref. 10 of Sect. IV]

$$p(\varphi) = |\langle \Psi | \phi \rangle|^2 = \frac{1}{2\pi} \frac{1}{1 + 2\langle N \rangle - 2\sqrt{\langle N \rangle(1 + \langle N \rangle)} \cos \varphi}$$

yielding that $\langle \varphi \rangle^2 \sim \frac{\pi^2}{8} \frac{1}{\langle N \rangle}$ and $\langle \varphi^2 \rangle \sim \frac{2 \log 2}{\langle N \rangle}$ such that the variance for large mean count $\lim_{\langle N \rangle \to \infty} \sigma_\varphi^2 \sim \frac{1}{\langle N \rangle} \to 0$.

Discussion of these results, comparison of different models proposed to solve these issues, and a detailed analysis are proposed in Sect. IV at the end of these notes.

I.2.8 Quasiprobability Representation of Operators

In the Glauber–$\mathcal{P}(\alpha)$ representation of operators [5], it is assumed that the density operator $\widehat{\Psi}$ obeys

$$\widehat{\Psi} = \int \mathcal{P}(\alpha, \alpha^*)|\alpha\rangle\langle\alpha|d(\Re e\,\alpha)d(\Im m\,\alpha) \qquad (I.2.22)$$

where $\int \mathcal{P}(\alpha, \alpha^*)d(\Re e\,\alpha)d(\Im m\,\alpha) = 1$. In this representation, the operators are normally ordered. In the case of antinormal ordering, a $\mathcal{Q}(\alpha, \alpha^*)$ representation can be defined..

Hence, $\mathrm{Tr}\big[\widehat{\Psi}\widehat{a}^{\dagger k}\widehat{a}^\ell\big] = \int \alpha^{*k}\alpha^\ell \mathcal{P}(\alpha, \alpha^*)d(\Re e\,\alpha)d(\Im m\,\alpha)$. Therefore $\mathcal{P}(\alpha, \alpha^*)$ can be interpreted as a probability density. This function must be positive and its derivative should be smooth. Its conditions of existence are sometimes difficult to verify as in the case of the real squeezed states [26]. To see this, let the $\mathcal{P}(\alpha)$ representation given by

$$\mathcal{P}_N(\alpha, \alpha^*) = \frac{1}{\pi^2} \int\!\!\int \chi_N(u, u^*) \exp(\alpha u^* - \alpha^* u)du\,du^* \qquad (I.2.23a)$$

and

$$\chi_N(u, u^*) = \mathrm{Tr}\left(\widehat{S}(\zeta)|0\rangle\langle 0|\widehat{S}^\dagger(\zeta)\exp(u\widehat{a}^\dagger)\exp(-u^*\widehat{a})\right) \qquad (I.2.23b)$$

be the characteristic function normally ordered.
The anti–normally ordered characteristic function $\chi_A(u, u^*)$ can be derived from $\chi_N(u, u^*)$ by using the Glauber identity (I.1.3) to obtain $\chi_A(u, u^*) = \exp\left(\frac{|u|^2}{2}\right)\chi_N(u, u^*)$.

Once again, by repeated use of (I.1.3), we easily find for $\alpha = 0$, $\phi = 0, \zeta = r$ in (I.2.23)

$$\chi_N(u, u^*) = \langle 0| \exp\left(u\widehat{A}_\zeta^\dagger\right) \exp\left(u^*\widehat{A}_\zeta\right) |0\rangle$$

$$= \exp\left(\frac{|u|^2}{2}\right) \langle 0| \exp\left(u\widehat{A}_\zeta^\dagger - u^*\widehat{A}_\zeta\right) |0\rangle$$

$$= \exp\left(-|u|^2 \sinh^2(r) + u_r \cosh(r)\sinh(r)\right)$$

where $u_r = \frac{u^2 + u^{*2}}{2}$. The Fourier transform diverges and the $\mathcal{P}$–representation cannot be defined for this type of state.

In fact, it is always possible to define a generalized positive $\mathcal{P}(\alpha, \beta)$ representation by extension of the dimension of the phase space so that a density operator can be written in the form [20, p. 408], $\left(d^2\alpha = d(\Re e\, \alpha)d(\Im m\, \alpha), \ d^2\beta = d(\Re e\, \beta)d(\Im m\, \beta)\right)$

$$\widehat{\rho} = \int \frac{1}{\langle \beta^*|\alpha\rangle} \mathcal{P}(\alpha, \alpha^*; \beta, \beta^*) |\alpha\rangle\langle\beta^*| \ d^2\alpha \ d^2\beta$$

this representation, being however not unique.

Now, instead of studying the unbounded operators $\widehat{p}$ and $\widehat{x}$, it is convenient, e.g. for the construction of the satisfactory phase operator, to consider the bounded functions $\left(\exp(\frac{i}{\hbar}p\widehat{x}), \ \exp(\frac{i}{\hbar}x\widehat{p})\right)$ which constitute the basis of the Weyl–representation of operators. With the Weyl unitary operator [9]

$$\widehat{W}(p, x) = \exp\left(i\left(p\widehat{x} - x\widehat{p}\right)\right) \tag{I.2.24}$$

we can write $|p, x\rangle = \widehat{W}(p, x)|0\rangle$, which is nothing but the phase–space rewriting of (I.2.1a). The normalization being given so that $\frac{1}{2\pi\hbar} \int |p, x\rangle\langle p, x| dp dx = \mathbb{1}$. We have

$$\begin{cases} \langle \alpha|\widehat{W}(p, x)|\beta\rangle = \int \alpha^*\left(\xi + \frac{x}{2}\right) \exp\left(ip\xi\right) \beta\left(\xi + \frac{x}{2}\right) d\xi \\ \langle p_2, x_2|p_1, x_1\rangle = e^{\frac{i}{2\hbar}(p_1 x_2 - x_1 p_2) - \frac{(p_2-p_1)^2 + \omega^2(x_2-x_1)^2}{4\hbar\omega}} \end{cases} \tag{I.2.25}$$

The reformulation of the coherent operators and states properties on the phase–state space can be extended to continuous representation

$$
\begin{cases}
\widehat{\Psi} = \displaystyle\int dp dx\, W(p,x)\widehat{\Delta}(p,x) \\[2mm]
W(p,x) = \displaystyle\int du\,\exp\left(\frac{i}{\hbar}xu\right)\langle p+\frac{u}{2}|\widehat{\Psi}|p-\frac{u}{2}\rangle = \mathrm{Tr}\left[\widehat{\Psi}\widehat{\Delta}(p,x)\right] \\[2mm]
\widehat{\Delta}(p,x) = \displaystyle\int dv\,\exp\left(\frac{i}{\hbar}pv\right)|x+\frac{v}{2}\rangle\langle x-\frac{v}{2}|
\end{cases}
$$

$$(\mathrm{I}.2.26a,b,c)$$

where $W(p,x)$ is the Wigner–function of the density operator $\widehat{\Psi}$. The function $W(p,x) \in \mathbb{R}$ that always exists, is normalized but not being positive–definite, cannot be interpreted as a probability density: it is a quasiprobability density. Its form for various states is given in [25]. Notice that $\langle x|\widehat{\Psi}|x\rangle = \int dp W(p,x)$. A similar relation holds for the integration over x. A useful formula for a product of two operators given by their Wigner–function $W_1(p,x), W_2(p,x)$ can be derived to obtain

$$
W_3(p,x) = \int \exp\left(2i\xi_1\zeta_2 - \xi_2\zeta_1\right) W_1(p+\zeta_1, x+\xi_1)\times
$$

$$
W_2(p+\zeta_2, x+\xi_2)d\xi_1 d\xi_2 d\zeta_1 d\zeta_2
$$

$$(\mathrm{I}.2.27)$$

This formula is sometimes utilized for rewriting the uncertainty relation of $\widehat{x},\widehat{p}$ and for rewriting a density operator

$$
W(p,x) = \gamma(p,x)^2 \left(1 - \frac{1}{32\xi_1^2\zeta_1^2}\left[2 - \left(\frac{x}{\xi_1}\right)^2 - \left(\frac{p}{\zeta_1}\right)^2\right]\right)
$$

$$(\mathrm{I}.2.28)$$

where $\gamma(p,x)^2 = \gamma_0 \exp\left(-\frac{x^2}{2\xi_1^2} - \frac{p^2}{2\varsigma_1^2}\right)$ is given.

By analogy with the classical definition, the quantum characteristic function can be written in the following form

$$\chi(u,v) = \mathrm{Tr}\left[\widehat{\Psi} \exp\left(\frac{i}{\hbar}(u\widehat{p} + v\widehat{x})\right)\right] \qquad (\mathrm{I.2.29})$$

A quantum characteristic operator can also be defined under the quantum stochastic rules (see subsection III.1.2). Taking the Fourier transform

$$\begin{cases} \mathcal{W}(p,x) = \mathrm{F.T.}\left[\chi(u,v)\right] \\ \qquad = \frac{1}{(2\pi\hbar)^2} \int\int du\,dv \exp\left(-\frac{i}{\hbar}(up + vx)\right)\chi(u,v) \\ \chi(u,v) = \int \exp\left(u\alpha^* - v\alpha\right)\mathcal{P}(\alpha,\alpha^*)d(\Re e\,\alpha)d(\Im m\,\alpha) \\ \qquad = \mathrm{Tr}\left[\widehat{\Psi}\exp(u\widehat{a}^\dagger)\exp(v\widehat{a})\right] \end{cases}$$

$$(\mathrm{I.2.30}a,b)$$

Several relations exist between these formulations.

For a squeezed state $|\alpha,\zeta\rangle$, the different representations $\mathcal{W}(p,x)$ can be shown to verify a convolution product

$$\mathcal{W}_{\alpha,\zeta}(p,x) = \widehat{\mathcal{W}}_{0,\zeta}(p,x) \star \mathcal{P}(p,x)$$
$$\mathcal{Q}(p,x) = \mathcal{W}_{0,\zeta i}(p,x) \star \mathcal{W}(p,x)$$

$$(\mathrm{I.2.31}a,b)$$

where $\mathcal{W}_{0,\zeta}(p,x) = \frac{1}{\pi}\exp\left(-\alpha_\zeta p^2 - \beta_\zeta x^2 + \gamma_\zeta px\right)$, the squeezing coefficients $\alpha_\zeta, \beta_\zeta, \gamma_\zeta$ being easily derived from (I.2.9e,f).

As a simple application, let us reproduce some elements of the demonstration of the quantum analog of the classical central limit theorem for a single mode of radiation [5].

Let the state $|z\rangle = |\frac{1}{\sqrt{M}} \sum_m z_m\rangle$ be composed of a large number M of independent identically distributed contributions z_m of distribution $r_m(z_m)$. Setting $d\mu(z_m) = \frac{1}{2\pi} d\Re(z_m) d\Im(z_m)$, the density operator is written

$$\hat{\Psi} = \int \mathcal{P}_M(z_m) \frac{1}{M} |\sum_m z_m\rangle\langle\sum_m z_m| d\mu(z_m)$$

where $\mathcal{P}_M(z_m)$ is given by a multiple convolution integral

$$\mathcal{P}_M(z_m) = \int \cdots \int r_1(z_m - z_1) r_2(z_1 - z_2)...r_M(z_{M-1})$$
$$d\mu(z_1)...d\mu(z_{M-1})$$

The Fourier transform $\mathcal{W}(x, k)$ of the equivalent form of $\mathcal{P}_M(z_m)$ in the phase space $W(p, q)$ is $\mathcal{W}(x, k) = \prod_{m=1}^{M} w_m\left(\frac{x}{\sqrt{M}}, \frac{k}{\sqrt{M}}\right)$ where $w_m(x, k)$ is the Fourier transform of the distribution $r_m(z)$ written in the phase space. From $|\mathcal{W}(x, k)| \leq \exp\left(\frac{1}{4}(x^2 + k^2)\right)$ yielding $\forall m$,

$$w_m(x, k) = w(x, k) \sim 1 - \frac{\langle|z|^2\rangle}{2}(x^2 + k^2)$$

For $M \to \infty$ and for small x, k, we obtain

$$\mathcal{W}(x, k) = \left(1 - \frac{\langle|z|^2\rangle}{2M}(x^2 + k^2)\right)^M \sim \exp\left(-\frac{\langle|z|^2\rangle}{2}(x^2 + k^2)\right)$$

$$W(p, q) = \frac{1}{\langle|z|^2\rangle} \exp\left(-\frac{1}{2\langle|z|^2\rangle}(p^2 + q^2)\right)$$

$$\mathcal{P}(z) = \frac{1}{\langle N\rangle} \exp\left(-\frac{1}{2\langle N\rangle}|z|^2\right)$$

$$(\text{I.2.31}c)$$

where we set $\langle N\rangle = \langle|z|^2\rangle$.

Therefore, using the expansion of $|z\rangle = \exp(-\frac{|z|^2}{2})\sum_n \frac{z^n}{\sqrt{n!}}|n\rangle$, we finally obtain

$$\hat{\Psi} = \frac{1}{\pi\langle N\rangle}\sum_{n,m=0}^{\infty}\frac{z^n z^{*m}}{\sqrt{n!m!}}|n\rangle\langle m|\left\{\int \exp\left(-|z|^2\left(1+\frac{1}{\langle N\rangle}\right)\right)d^2z\right\}$$

$$= \frac{1}{\langle N\rangle}\sum_{n=0}^{\infty}\frac{1}{n!}\left\{\int |z|^{2n}\exp\left(-|z|^2\left(1+\frac{1}{\langle N\rangle}\right)\right)d|z|^2\right\}|n\rangle\langle n|$$

$$= \frac{1}{\langle N\rangle}\sum_{n=0}^{\infty}\left(1+\frac{1}{\langle N\rangle}\right)^{-(n+1)}|n\rangle\langle n| \equiv (1-v)\sum_{n=0}^{\infty}v^n|n\rangle\langle n|$$

$$(I.2.31d)$$

which is the geometric form of a thermal state with the notations $v = \frac{\langle N\rangle}{1+\langle N\rangle} \equiv e^{-w} \equiv e^{-\frac{\hbar\omega}{kT}}$. This density operator can now be expanded in terms of the number operator $\hat{N} = \hat{a}^{\dagger}\hat{a}$ and in the $P(\alpha)$–representation

$$\hat{\Psi} = \left(1-e^{-w}\right)e^{-w\hat{N}} = \frac{1}{\pi\langle N\rangle}\int exp\left(-\frac{|\alpha|^2}{\langle N\rangle}\right)|\alpha\rangle\langle\alpha|d^2\alpha$$

If we now superimpose a coherent state of amplitude μ, the resulting density operator will be in the form used in the following

$$\hat{\Psi}(\mu) = D(\mu)\hat{\Psi}(0)D^{\dagger}(\mu)$$

$$= \left(1-e^{-w}\right)\exp\left(-w(\hat{a}^{\dagger}-\mu^*)(\hat{a}-\mu)\right)$$

$$= \frac{1}{\pi\langle N\rangle}\int exp\left(-\frac{|\alpha-\mu|^2}{\langle N\rangle}\right)|\alpha\rangle\langle\alpha|d^2\alpha$$

An important example, pointed out by Glauber, concerns the case for which

$$\widehat{\Psi} = (1+\mu)\widehat{\Psi}_t - \mu \sum_n \delta(n)|n\rangle\langle n|$$

where $\widehat{\Psi}_t$ is given by (I.2.31d).

The condition $(\langle n|\widehat{\Psi}|n\rangle \geq 0, \ \forall n)$ required in order that $\widehat{\Psi}$ is a density operator, reads $-1 \leq \mu \leq \frac{1}{\langle N \rangle}$. Moreover, the reduced variance of the photon number $\sigma_n^2 - \langle n \rangle = (1 - \mu^2)\langle N \rangle^2$ shows that an antibunching effect may occur only for $\mu > 1 \Longleftrightarrow \langle N \rangle < 1$.

I.3 Coherent Receiver Operators

Let the field to be detected, be a superposition, via a beam splitter of transmission coefficient $\sqrt{\epsilon}$, of a source field ($\widehat{a}, \widehat{a}^\dagger$, frequency ω_s), in a state defined by the density operator $\widehat{\rho}_a(\alpha)$. A local oscillator (LO) defined by ($\widehat{b}, \widehat{b}^\dagger$, frequency ω_0), is in a state defined by the operator $\widehat{\rho}_b(\beta) = |\beta\rangle\langle\beta|$, where we assume that $\beta \in \mathbb{R}$, for simplificity. We denote ω_i the frequency of the image field, such that $\omega_s = \omega_0 + \omega_i$.

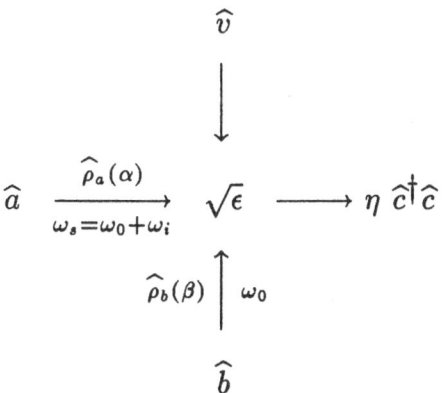

A vacuum field $\widehat{v} = \widehat{v}_1 + i\widehat{v}_2$ is supposed to enter the fourth port of the receiver. At the output, a photodetector of efficiency η measures the field ($\widehat{c}, \widehat{c}^\dagger$).

I.3.1 Homodyne

Here $\omega_i = 0$. The homodyne measurement operator can be demonstrated to be

$$\widehat{a_1} = \frac{1}{2}\left(\hat{a} + \hat{a}^\dagger\right) \tag{I.3.1}$$

We reproduce part of the calculations first demonstrated in [14]. The field $\hat{c} = \sqrt{\epsilon}\hat{a} + \sqrt{1-\epsilon}\hat{b}$ in the state defined by the density operator $\hat{\rho}_c(\alpha, \beta; \epsilon)$ is detected by a photodetector of efficiency η. The characteristic function of the measured operator

$$\widehat{M} = \frac{1}{\beta}\left(\hat{c}^\dagger\hat{c} - \epsilon\eta\,\mathrm{Tr}\,\left(\hat{\rho}_a(\alpha)\hat{a}^\dagger\hat{a}\right) - \epsilon(1-\epsilon)\beta^2\right)$$

takes the form

$$\Phi_{\widehat{M}}(u) = \mathrm{Tr}(\hat{\rho}_c\widehat{M})$$

$$= \left(\frac{1}{1-q}\int e^{-\frac{1-q}{1-\epsilon+q\epsilon}\left|\beta\sqrt{1-\epsilon}+\sqrt{\epsilon}\alpha\right|^2}\hat{\rho}_a(\alpha)\frac{d^2\alpha}{\pi}\right)\psi(u)$$

where $q = 1 - \eta + \eta\exp\left(\frac{iu}{\beta}\right)$ and

$$\psi(u) = \exp\left(-\frac{iu}{\beta}\left(\eta(1-\epsilon)\beta^2 + \eta\epsilon\,\mathrm{Tr}\,\left(\hat{\rho}_a\hat{a}\hat{a}^\dagger\right)\right)\right)$$

For a LO of very high intensity level, it can be proved that

$$\lim_{\beta\to\infty}\Phi_{\widehat{M}}(u) = e^{-\frac{\eta(1-\epsilon)(1-2\epsilon\eta)}{2}u^2}\phi_{\widehat{\rho_a}}\left(0, u\eta\sqrt{\epsilon(1-\epsilon)}\right)$$

$$= \exp\left(\frac{u^2}{8}\right)\phi_{\widehat{\rho_a}}\left(0, \frac{u}{2}\right)$$

(I.3.2)

where

$$\phi_{\widehat{\rho}_a}(u) = \text{Tr}\left(\widehat{\rho}_a \exp\left(-u^*\widehat{a}\right) \exp\left(u\widehat{a}^\dagger\right)\right) = e^{-\frac{|u|^2}{2}} \text{Tr}\left(\widehat{\rho}_a \widehat{D}(u)\right)$$

(I.3.3)

$u = u_1 + iu_2$ and $\widehat{D}(u)$ is the displacement operator (see § I.2.2). To demonstrate the results, it is first noticed

$$\text{Tr}\left(q^{\widehat{c}^\dagger\widehat{c}} \exp(-u\widehat{c}^\dagger) \exp(u^*\widehat{c})\right)$$

$$= \frac{1}{\pi} \exp\left(|u|^2\right) \sum_{j,k} \int d^2\gamma \exp\left(u^*\gamma - u\gamma^*\right) \exp\left(-|\gamma|^2\right) \frac{\gamma^j \gamma^{*k}}{\sqrt{j!k!}} q^j q^{*k} \langle k|j\rangle$$

$$= \frac{1}{1-q} \exp\left(-\frac{q}{1-q}|u|^2\right)$$

It is then proved that the operator $\widehat{M'} = \frac{\widehat{M}}{2\sqrt{\eta\epsilon(1-\epsilon)}}$, is such that

$$\lim_{\substack{\beta\to\infty \\ \epsilon\to 1}} \Phi_{\widehat{M'}}(u) = \exp\left(-\frac{1-2\eta}{8}u^2\right) \phi_{\widehat{\rho}_a}\left(0, \frac{u\sqrt{\eta}}{2}\right)$$

(I.3.4)

which is interpreted as the characteristic function of the operator $\widehat{a'_1} = \sqrt{\eta}\widehat{a_1} + \sqrt{1-\eta}\widehat{v_1}$ for $\eta \to 1$, where $\widehat{v_1}$ is defined such that $\widehat{v}|0\rangle \equiv (\widehat{v_1} + i\widehat{v_2})|0\rangle = 0$.

Remark: This result means that the photodetector works as a classical device ($\eta = 1$) provided that a vacuum field is added to the signal field via the beam splitter.

Finally, for $u \in \mathbb{R}$, and

$$\widehat{\rho}_a(\beta) = \frac{1}{\pi N_1} \int \exp\left(-\frac{|\alpha - \beta|^2}{N_1}\right) |\alpha\rangle\langle\alpha| d^2\alpha$$

(I.3.5)

eq. (I.3.3) leads to

$$\Phi_{\widehat{a_1}}(u) = \exp\left(u\,\Re e(\beta)\right)\exp\left(-\frac{u^2}{2}\left(\frac{N_1}{2}+\frac{1}{4}\right)\right) \qquad (\text{I.3.6})$$

which is typical of a Gaussian random variable X: $\Phi_X(u) = \exp\left(iu\langle X\rangle - \frac{\sigma_X^2 u^2}{2}\right)$.

I.3.2 Heterodyne

Here $\omega_i \neq 0$. The heterodyne measurement operator can be shown to be

$$\widehat{a_2} = \frac{1}{2}\left(\widehat{a} + \widehat{a_i}^\dagger\right) \qquad (\text{I.3.7})$$

Details of the calculations are similar to those given in the preceeding subsection. The field to be considered is

$$\begin{cases} \widehat{c} = \sqrt{\eta}\,\widehat{c_0} + \sqrt{1-\eta}\left(\widehat{v_1}e^{-i(\omega_0+\omega_i)t} + \widehat{v_2}e^{-i(\omega_0-\omega_i)t}\right) \\ \widehat{c_0} = \sqrt{\epsilon}\left(\widehat{a}e^{-i(\omega_0+\omega_i)t} + \widehat{a_i}e^{-i(\omega_0-\omega_i)t}\right) + \sqrt{1-\epsilon}\,\widehat{b}\,e^{-i\omega_0 t} \end{cases}$$

$$(\text{I.3.8})$$

Let us apply these results in the limit of $\beta \to \infty$, to a specific example that is very useful in optical communication. Let the density operators of the signal and image fields be given by

$$\begin{cases} \widehat{\rho_a}(\beta) = \frac{1}{\pi N_1}\int \exp\left(-\frac{|\alpha-\beta|^2}{N_1}\right)|\alpha\rangle\langle\alpha|d^2\alpha \\ \widehat{\rho_i} = \frac{1}{\pi N_2}\int \exp\left(-\frac{|\alpha|^2}{N_2}\right)|\alpha\rangle\langle\alpha|d^2\alpha \end{cases} \qquad (\text{I.3.9}a,b)$$

Using the signal–image independence, the operators

$$\begin{cases} \widehat{L_1} = \frac{1}{2}\left((\widehat{a_s}+\widehat{a_s}^\dagger)+(\widehat{a_i}+\widehat{a_i}^\dagger)\right) = \widehat{a_1}+\widehat{a_2} \\ \widehat{L_2} = \frac{1}{2i}\left((\widehat{a_s}-\widehat{a_s}^\dagger)-(\widehat{a_i}-\widehat{a_i}^\dagger)\right) \end{cases} \qquad (\text{I.3.10}a,b)$$

are commuting. The joint characteristic function is found to be of Gaussian shape

$$\Phi_{\widehat{L}_1,\widehat{L}_2} = \mathrm{Tr}\left(\widehat{\rho}_a\widehat{\rho}_i \exp\left(u\widehat{L}_1\right)\exp\left(v\widehat{L}_2\right)\right)$$

$$= \exp\left(u\,\Re e\left(\alpha_s\right) + v\,\Im m\left(\alpha_s\right)\right)\exp\left(-\frac{N_1 + N_2 + 1}{2}\left(\frac{u^2 + v^2}{2}\right)\right)$$

$$(\mathrm{I.3.11})$$

where $u, v \in \mathbb{R}$. In the case that $N_2 \ll N_1$, eq. (I.3.11) is simplified so that $\sigma_H^2 = \frac{N_1}{2} + \frac{1}{2}$ which is larger than the value found in the case of homodyne detection where $\sigma_h^2 = \frac{N_1}{2} + \frac{1}{4}$ (see eq. (I.3.5)).

Another measurement scheme has recently been extensively studied because it provides new interesting applications in quantum optical communication. It is called *quantum non demolition* measurement because the observable is measured without any perturbation.

I.3.3 Quantum Non Demolition

Suppose the process $\widehat{X}(t)$ is observable by the measurement of $\widehat{Y}(r)$ so that the commutation relation $[\widehat{X}(s), \widehat{Y}(t)] = 0$, $\forall t \geq s$ without any restriction on the choice of the observables measured at future time instants,

$$[\widehat{Y}(t), \widehat{Y}(u)] = 0, \qquad u \geq t$$

After the measurement at time s, the state of $\widehat{X}(t)$ is conserved at all times. Such a measurement is called of *Quantum Non-Demolition* (QND) type [34]. The main consequence is that the back action in the process may be evaded, therefore repeated measurements become possible.

Experimental realisations have been reported using a set up of the type of Fig. 0.3. using a lossless Kerr effect characterized by an intercation Hamiltonian $\widehat{H_I} = \chi(\omega_s, \omega_p)\widehat{N_s}\widehat{N_p}$. The required conditions are for $\widehat{N_s}$ the measured observable, $[\widehat{H_I}, \widehat{N_s}] = 0$ (back–action evasion), $[\widehat{N_s}(0), \widehat{N_s}(t)] = 0$ (QND measurement) and the $\widehat{S_p}$ read-out observable (sine operator of the probe field) $[\widehat{H_I}, \widehat{S_p}] \neq 0$ [15]. Application of QND to optical communication is briefly examined in an example treated below.

Consider a QND transmission scheme where we denote $\widehat{X} \equiv \widehat{X_1^s} = \frac{1}{2}(\widehat{a_s} + \widehat{a_s}^\dagger)$ of eigenstates $|x_1\rangle$ and $\widehat{Y} \equiv \widehat{X_2^p} = \frac{1}{2i}(\widehat{a_p} - \widehat{a_p}^\dagger)$ of eigenstates $|x_2\rangle$. The evolution operator $\widehat{U}(0,t)$ of the system (a signal with a probe), characterized by the interaction Hamiltonian $\widehat{H_I}(\theta) = \hbar\chi\widehat{X_1^s}\widehat{X_2^p}$, is given by

$$\widehat{U}(s,t) = \exp\left(-i\chi\int_s^t \widehat{H_I}(\theta)d\theta\right)$$

Initially, we suppose that the signal and probe are independent and prepared in squeezed states so that $|\phi_s\rangle : (\langle x_1\rangle, \sigma_s^2)$, $|\eta_p\rangle : (\langle x_2\rangle = 0, \sigma_p^2)$. At time t, the density operator is given by $\Lambda(t) = |\psi\rangle\langle\psi|$, where $|\psi\rangle = |\phi_s\rangle|\eta_p\rangle$.

To evaluate $\mathrm{Prob}(x_1, x_2; t)$ the probability distribution of outcomes $\{x_1, x_2\}$, we employ the evolution equation in the interaction picture, where we set $\widehat{\Pi}_k^{(a)} \equiv |x_1\rangle\langle x_1|$ and $\widehat{\Pi}_q^{(b)} \equiv |x_2\rangle\langle x_2|$. It is easily demonstrated that

$$\mathrm{Prob}(x_1, x_2; t) = \mathcal{N}_{x_1}(\langle x_1\rangle, \sigma_s^2)\mathcal{N}_{x_2}(2\chi t\langle x_1\rangle, \sigma_p^2)$$

where $\mathcal{N}_X(m, \sigma^2)$ denotes a Gaussian distribution of the r.v. of mean m and variance σ^2 [33]. The marginal distribution for the probe field is $\mathcal{N}_{x_2}(2\chi t\langle x_1\rangle, (2\chi t)^2\sigma_s^2 + \sigma_p^2)$. It is also noticed that for $\chi t \to \infty$, $\mathrm{Prob}(x_1|x_2; t) \to \mathcal{N}_{x_1}(\frac{x_2}{2\chi t}, \frac{\sigma_p}{2\chi t})$ which means that after the outcome x_2 is registered, the variance of the signal observable is significantly decreased.

An interesting situation occurs when the intensity of the probe is large enough. In the case of photon number measurements (the readout observable becomes $\widehat{S_p} = \sin(\widehat{\Phi_p})$), the measurement accuracy is proportional to the variance of the phase of the probe field. Therefore a definition of the phase operator is absolutely required. This question is analyzed in some detail in Sect. IV.

I.4 Measurement Processes

One of the problems currently discussed in the theory of measurements in quantum mechanics, is related to the *wave-packet reduction postulate* (WRP). According to this postulate, after a measurement of an observable of the system by an apparatus, the system evolves from a mixture of states of positive entropy to a pure state of zero entropy. This apparently contradicts the classical thermodynamics laws. Moreover the information transfer from the system to the apparatus is an irreversible process.

In fact, to understand the physics of the measurement, the process should be examined on the basis of whether or not the measurement is read. This is considered in the following section.

I.4.1 Time Evolution

Let a system S be given on which the observable $\mathcal{A}$ is characterized by the self–adjoint linear operator on $\mathcal{H}$, such that $\widehat{A} = \sum_k a_k \widehat{\Pi}_k^{(a)}$. We set $|a_k\rangle\langle a_k| \equiv \widehat{\Pi}_k^{(a)}$ with the orthogonal resolution of the identity

$$\begin{cases} \widehat{\Pi}_k^{(a)} \widehat{\Pi}_\ell^{(a)} = 0 \\ \sum_k \widehat{\Pi}_k^{(a)} = \mathbb{1} \end{cases} \qquad (I.4.1)$$

Let $\widehat{\Lambda}$ be the density operator of the system at the initial time s. Consider now this quantum system S coupled with $\mathcal{I}$, an apparatus (or instrument) initially separated from it. This apparatus, utilized to provide information about S, is in interaction with S under the self–adjoint Hamiltonian $\widehat{H} = \widehat{H}_0 + \widehat{H}_{SI}$, $\widehat{H}_0$ being the Hamiltonian of the free system.

The system evolves with a unitary operator $\widehat{U}(s,t)$ following the *evolution postulate*, and obeys (the model is simplified by considering that all operators act on the same Hilbert space $\mathcal{H}$)

$$\begin{cases} \dfrac{d}{dt}\widehat{\Lambda}(t) = -\dfrac{i}{\hbar}\left[\widehat{H},\widehat{\Lambda}\right] \\ \widehat{\Lambda}(t) = \widehat{U}(s,t)\widehat{\Lambda}(s)\widehat{U}^\dagger(s,t) \\ \widehat{U}(s,t) = \exp\left(-\dfrac{i}{\hbar}\int_s^t \widehat{H}(\vartheta)d\vartheta\right) \end{cases} \qquad (I.4.2)$$

To simplify the presentation of this section, only a few results of the theory of quantum measurements established by von Neumann are utilized here. A point of view suitable for information communication purposes is adopted here.

For a single measurement and after an interaction at t, $\mathcal{S}$ and $\mathcal{I}$ are separated. The outcome a_k is registered with the probability according to the *fundamental postulate* (postulate of observables)

$$\text{Prob}\,(a_k) = \text{Tr}\left[\widehat{\Lambda}(t)\widehat{\Pi}_k^{(a)}\right] \qquad (I.4.3)$$

Its average value is given by

$$\text{E}[a_k] = \sum_k a_k \,\text{Prob}\,(a_k) = \sum_k \text{Tr}\left(a_k \widehat{\Pi}_k^{(a)}\right)\widehat{\Lambda}(t) = \text{Tr}\left(\widehat{A}\widehat{\Lambda}(t)\right)$$

After such an observation, the density of the system becomes by the WRP:

$$\widehat{\Lambda}'(t) = \frac{\widehat{\Pi}_k^{(a)}\,\widehat{\Lambda}(t)\widehat{\Pi}_k^{(a)}}{\text{Tr}\left[\widehat{\Lambda}(t)\widehat{\Pi}_k^{(a)}\right]}$$

Notice that if the measurement is not registered, the final density is

$$\widehat{\Lambda}''(t) = \sum_k \widehat{\Lambda}'(t)\,\text{Tr}\left[\widehat{\Lambda}(t)\widehat{\Pi}_k^{(a)}\right] = \sum_k \widehat{\Pi}_k^{(a)}\,\widehat{\Lambda}(t)\widehat{\Pi}_k^{(a)}$$

Initially in the state $|\psi(0)\rangle = |a_k\rangle \otimes |\varphi\rangle$, the global system $(\mathcal{S}+\mathcal{I}^{(a)})$ evolves to the state $|\psi(t) =\rangle|a_k\rangle \otimes |i_k^{(a)}\rangle$ for which the readout k ($|i_k^{(a)}\rangle$) on the apparatus means that the system $\mathcal{S}$ is in $|a_k\rangle$. This type of measurement is called *ideal*.

Now according to the *superposition* principle if we start with $|\psi(0)\rangle = \sum_k p_k |a_k\rangle \otimes |\varphi\rangle$, we end up with $|\psi(t)\rangle = \sum_k p_k |a_k\rangle \otimes |i_k^{(a)}\rangle$ for which quantum correlations appear between the macroscopic states $|a_k\rangle \otimes |i_k^{(a)}\rangle$ and $|a_l\rangle \otimes |i_\ell^{(a)}\rangle$. Moreover, it is possible to set $\sum_k |a_k\rangle \otimes |i_k^{(a)}\rangle \equiv \sum_m q_m |b_m\rangle \otimes |i_m^{(a)}\rangle$ which means that the apparatus is measuring a different observable $\widehat{B}$ of $\mathcal{S}$.

Several physicists propose to utilize this ambiguity to encrypt information for secure transmission.

A slightly more general concept is perhaps more suitable to define a generalized observable. Let $\hat{\sigma}$ on $\mathcal{K}$ be the initial density operator of the instrument $\mathcal{I}$ with a *pointer* observable $\widehat{I}_k^{(a)}$. The interaction with the system of density operator $\hat{\Lambda}$ occurs at the time t. The fundamental postulate, also called the postulate of *reproducibility* is

$$\mathrm{Tr}\left[\left(\mathbb{1} \otimes \widehat{I}_k^{(a)}\right)\left(\hat{U}(s,t)(\hat{\Lambda} \otimes \hat{\sigma})\hat{U}(s,t)^{\dagger}(t)\right)\right] = \mathrm{Tr}\left[\hat{\Lambda}\widehat{\Psi}_k^{(a)}\right]$$

The set $\left\{\widehat{\Psi}_k^{(a)}\right\}$ need not be orthogonal. Therefore, we choose them so that $\widehat{\Psi}_\ell^{(a)}|a_k\rangle = \psi_{k\ell}|a_k\rangle$. Now, from $\sum_k \widehat{\Psi}_\ell^{(a)} \widehat{\Pi}_k^{(a)} = \sum_k \psi_{k\ell}\widehat{\Pi}_k^{(a)}$, we have the following properties for the non $\perp$ operators

$$\begin{cases} \widehat{\Psi}_\ell^{(a)} = \sum_k \psi_{k,\ell}\widehat{\Psi}_k^{(a)} \\ \widehat{\Psi}_\ell^{(a)} \widehat{\Psi}_m^{(a)} \neq \delta_{\ell,m} \\ [\widehat{\Psi}_\ell^{(a)}, \widehat{\Psi}_m^{(a)}] = 0 \\ \sum_\ell \widehat{\Psi}_\ell^{(a)} = \mathbb{1} \\ \mathrm{Tr}\,\widehat{\Psi}_\ell^{(a)} = 1 \end{cases}$$

Let us extend this formalism to the case of repeated measurements made by two different apparatus $\mathcal{I}_a, \mathcal{I}_b$ on $\hat{A}$ and $\hat{B}$ at t and t'. After the measurement a_k is registered, the system is in $\widehat{\Lambda}'(t)$. At t', we have $\widehat{\Lambda}(t') = \hat{U}(t,t')\widehat{\Lambda}'(t)\hat{U}^{\dagger}(t,t')$. Therefore, using the WRP, we obtain the conditional measurement probability

$$\text{Prob}(b_q|a_k) = \text{Tr}\left[\widehat{\Lambda}(t')\widehat{\Pi}_q^{(b)}\right]$$

$$= \text{Tr}\left[\widehat{U}(t,t')\left(\frac{\widehat{\Pi}_k^{(a)}\widehat{\Lambda}(t)\widehat{\Pi}_k^{(a)}}{\text{Tr}\left[\widehat{\Lambda}(t)\widehat{\Pi}_k^{(a)}\right]}\right)\widehat{U}^\dagger(t,t')\widehat{\Pi}_q^{(b)}\right]$$

$$(\text{I.4.4}a)$$

The joint distribution being given by

$$\text{Prob}(a_k, b_q) = \text{Prob}(b_q|a_k)\,\text{Prob}(a_k)$$

$$= \text{Tr}\left[\widehat{U}(t,t')\widehat{\Pi}_k^{(a)}\widehat{\Lambda}(t)\widehat{\Pi}_k^{(a)}\widehat{U}^\dagger(t,t')\widehat{\Pi}_q^{(b)}\right] \qquad (\text{I.4.4}b)$$

As demonstrated in [31], if we treat the whole system $\mathcal{S} + \mathcal{I}_a + \mathcal{I}_b$, only the first postulate of quantum mechanics is necessary in order to obtain equations (I.4.4a,b).

An important question of current interest can be stated as follows: the apparatus being quantum mechanically described, what is the process leading to the selection of the right p.o.m. among the variety $\widehat{\Pi}_k^{(a)}, \widehat{\Pi}_q^{(b)}, \ldots$?

An answer to this question was recently developed [18,19] suggesting that the apparatus $\mathcal{I}$ is actually in interaction with the environment $\mathcal{E}$ via the Hamiltonian $\widehat{H}_{\mathcal{IE}}$.

In case such interaction dominates (the system $\mathcal{S}$ that is microscopic is not directly in interaction with $\mathcal{E}$ that is macroscopic) and assuming that $\widehat{H}_{\mathcal{IE}}$ and $\widehat{H}_{\mathcal{E}}$ are diagonal in the same normalized basis $\{|i_n\rangle\}$, the final state after measurement can be written as $|\psi_f(t)\rangle = \sum_n c_n |s_n(t)\rangle \otimes |i_n(t)\rangle \otimes |e_n(t)\rangle$ for which we notice that there is, at first, a measurement of $\mathcal{S}$ by $\mathcal{I}$, and after a QND measurement of $\mathcal{I}$ by $\mathcal{E}$. This latter is an ideal one as for $t \to \infty$, the environment states become uncorrelated $\langle e_m(t)|e_n(t)\rangle \to \delta_{mn}$.

Let us briefly see how the entropy functions evolve during these steps.

I.4.2 Entropy Evolution

We adopt the conventional interpretation of the concept of entropy in the sense that it is not an observable, i.e. an operator on Hilbert space but rather a property specifying the state of the system described by the density matrix $\hat{\rho}$. In fact, a measurement which is basically a transfer of information is always incomplete and the lack of knowledge from the maximum of available information can be characterized by the entropy.

Among the different definitions of the entropy, we select the von Neumann entropy that satisfyes the basic mathematical conditions and the statistical physics requirements

$$S(\hat{\rho}) = -\mathrm{Tr}\ (\hat{\rho}\log\hat{\rho}) \tag{I.4.5a}$$

Notice that for a system having a given Hamiltonian $\hat{H}$, the entropy defined by (I.4.5a) is such that $\frac{\partial}{\partial t}S = -\mathrm{Tr}\ \left(\hat{H}.\left[\hat{\rho},\log\hat{\rho}\right]\right) = 0$, where $\frac{\partial}{\partial t}\hat{\rho} = \left[\hat{\rho},\hat{H}\right]$. Similarly, the relative entropy is

$$S(\hat{\Lambda}\|\hat{\rho}) = -\mathrm{Tr}\ \hat{\rho}\left(\log\hat{\rho} - \log\hat{\Lambda}\right) \tag{I.4.5b}$$

whose the properties of positivity, convexity and the usual inequalities can be demonstrated. We just mention here, in a a simplified presentation, a very few of the results which are of direct relevance with the subject of these lecture notes. We refer to the book by Thirring for details and for the proofs of the results mentioned here [17].

As known, it is often possible to associate a classical density with a density operator $\hat{\rho}$ in the form $p_c(\beta) = \langle \beta|\hat{\rho}|\beta\rangle$ that must be positive and normalized (β could be a coherent state). Therefore its differential entropy takes the form

$$S(p_c) = -\int p_c(\beta) \log p_c(\beta) d^2\beta$$

Now in the case that the density operator can be expanded in the $P(\alpha)$–representation (see § I.2.8), another differential entropy can be introduced by using the usual form

$$S(P) = -\int P(\alpha) \log P(\alpha) d^2\alpha$$

Then the following inequalities hold

$$0 \leq S(P) \leq S(\hat{\rho}) \leq S(p_c) \tag{I.4.5c}$$

Finally, for $\lambda_i > 0, \sum_i \lambda_i = 1$, it can be shown that

$$S\left(\sum_i \lambda_i \hat{\rho}_i\right) \leq \sum_i \lambda_i S(\hat{\rho}_i) - \sum_i \lambda_i \log \lambda_i \tag{I.4.5d}$$

After a measurement a_k (that could be degenerate) with probability p_k of an observable $\hat{A} = \sum_k a_k \widehat{\Pi}_k^{(a)}$ of the system S by an apparatus $\mathcal{I}$, the system evolves from $\hat{\rho}$, a mixture of states of entropy $S(\hat{\rho}) \geq 0$ to $\widehat{\rho_k}$, (generally not a pure state) of entropy $S(\widehat{\rho_k})$. The process should be examined on the basis of whether or not the measurement is read and divided into stages such that

$$\hat{\rho} \quad \rightarrow \quad \hat{\rho}' \quad \rightarrow \quad \widehat{\rho_k} \tag{I.4.6}$$

for which we have

$$\begin{cases} S(\widehat{\rho'}) \geq S(\widehat{\rho}) \\ S(\widehat{\rho'}) \geq \langle S(\widehat{\rho_k}) \rangle \end{cases} \qquad (I.4.7a,b)$$

where $\langle S(\widehat{\rho_k}) \rangle = -\sum_k p_k \text{Tr} \left(\widehat{\rho_k} \log \widehat{\rho_k} \right)$ and $\widehat{\rho_k} = \dfrac{\widehat{\Pi_k}^{(a)} \widehat{\rho} \widehat{\Pi_k}^{(a)}}{p_k}$.

In fact, the measurement stage of the scheme (I.4.6), which is a deterministic one, yields no information. This is because it implies quantum correlations due to the interaction $S - I$ which vanish as soon as the system and the apparatus are separated. During this stage, irreversible processes take place as a result of an energy transfer that affects the system under observation (e.g. quantum Zenon effect). Notice that although an apparatus is a collection of quantum microscopic elements, this macroscopic object must be deterministic for the outcomes to be reproducible. The reading (undeterministic process with no time evolution equation) which is a selection among all systems, informs us about $\widehat{\rho'}$ instead of $\widehat{\rho}$. It allows us to gain some information as seen in I.4.7b.

I.5 References

[1]   A. Messiah: *Mécanique Quantique* (Dunod, Paris 1959).

[2]   I.I. Gol'dman and al: *Problems in Quantum Mechanics*, ed. by D. ter Haar (Infosearch Ltd, London 1960).

[3]   R.J. Glauber: *Quantum Optics and Quantum Electronics*, ed. by C. de Witt, A. Blandin and C. Cohen–Tannoudji (Gordon and Breach, New York 1964).

[4]   W.H. Louisell: *Radiation and Noise in Quantum Electronics* (Mc Graw Hill Book Company, New York 1964).

[5]   J.R. Klauder, E.C.G. Sudarshan: *Fundamental of Quantum Optics* (Benjamin, New York 1968).

[6]   P. Carruthers, M. M. Nieto: Rev. of Mod. Phys., **40** 411 (1968).

[7]   T.W.B. Kibble: *Quantum Optics*, ed. by S.M. Kay and A. Maitland (Academic Press, New York 1970).

[8]   R. Loudon: *The quantum theory of light* (Clarendon Press, Oxford 1973).

[9]   S.R. de Groot: *La transformation de Weyl et la fonction de Wigner: une forme alternative de la mécanique quantique* (Presses de l'Université de Montréal 1974).

[10]   C.W. Helstrom: *Quantum Detection and Estimation Theory* (Academic Press, New York 1976)

[11]   T.S. Santhanam, A.R. Tekumalla: Found. of Physics, **6** 583 (1976).

[12]   H.P. Yuen: Phys. Rev. A, **13** 2226 (1976).

[13]   C. Cohen–Tannoudji, B. Diu, F. Laloë: *Mécanique Quantique* (Hermann, Paris 1977) part I.

[14]   H.P Yuen, J.H. Shapiro: in *Coherence and Quantum Optics IV*, ed. by L. Mandel and E. Wolf (Plenum Press, New York 1978).

[15]   C.M. Caves, K.S. Thorne, R.W. Drever, V.D. Sandberg and M. Zimmermann: Rev. Mod. Phys., **52** 341 (1980).

[16]   C.M. Caves: Phys. Rev. D, **23** 1693 (1981).

[17]   W. Thirring: *Quantum mechanics of Atoms and Molecules*. A course in Mathematical Physics 3, (Springer–Verlag, Wien 1981).

[18]   W. Zurek: Phys. Rev. D, **24** 1516 (1981).

[19]   W. Zurek: Phys. Rev. D, **26** 1862 (1982).

[20]   C.W. Gardiner: *Handbook of Stochastic Methods* (Springer–Verlag, Berlin 1983).

[21]   A.M. Perelomov: *Generalized coherent states and their applications* (Springer–Verlag, Berlin 1986).

[22]   S. M. Barnett, D. T. Pegg: Jour. Phys. A: Math. Gen., **19** 3849 (1986).

[23]   K. Kraus: Phys. Rev. D, **35** 3070 (1987).

[24]   S.M. Barnett, P.L. Knight: Jour. of Modern Optics, **34** 841 (1987).

[25]   G.S. Agarwal: Jour. of Modern Optics, **34** 909 (1987).

[26]   M. Hillery: Phys. Rev. A, **35** 725 (1987).

[27]   R. Loudon, P.L. Knight: Jour. of Modern Optics, **34** 709 (1987).

[28]   H. Maassen, J.B.M. Uffink: Phys. Rev. Lett., **60** 1103 (1988).

[29]   J. Rai, C.L. Mehta: Phys. Rev. A, **37** 4497 (1988).

[30]   H. Fearn, M.J. Collett: Jour. of Modern Optics, **35** 553 (1988).

[31]   C. Cohen–Tannoudji: *Cohérences quantiques et dissipation*, Cours du Collège de France, Paris 1989–1990.

[32]   S. Reynaud: Ann. Phys. Fr., **15** 63 (1990).

[33]   T. Uyematsu, O. Hirota, K. Sakaniwa: *Quantum Aspects of Optical Communications*, ed. by C. Bendjaballah, O. Hirota and S. Reynaud (Springer Verlag, Berlin 1991).

[34]   V.B. Braginsky, F. Ya. Khalili: *Quantum measurement*, (Cambridge Univ. Press, New York 1992).

[35]   C.L. Mehta, A.K. Roy, G.M. Saxena: Phys. Rev. A, **46** 1565 (1992).

[36]   S. Abe, R. Ehrhardt: Phys. Rev. A, **48** 986 (1993).

[37]   L.P. Hughston, R. Jozsa, W.K. Wootters: Phys. Lett. A, **183** 14 (1993).

II. Performance Criteria: Detection, Information

This section deals with the criteria of the statistical detection and information theories. Basic results of both formulations, classical as well as quantum are recalled and demonstrated when it is necessary. Applications to cases of practical importance are treated in detail.

II.1 Classical Formulation

Once the receiver operator is fixed, the system performances can be calculated using the results of the classical theory of detection and estimation for detection purposes.

On the other hand, as the information criterion is closely related to entropy, before applying the theory to optical communications, it is useful to set out the main properties of the entropy function in terms of classical probability.

We begin with the presentation of some elements of the theory of statistical detection.

II.1.1 Statistical Detection

The standard theory shows that detection optimality is attained using the maximum likelihood ratio test.

In this subsection, we propose to show some calculation of this test for photon counting processing in the case of PPM modulation format.

II.1.1a Likelihood Ratio Test

Let Y be the M–dimensional vector of the M photocounts $\{N_i\}$, $(i = 1, ..., M)$. The conditional probability distribution $p(y|m)$, $m = 1, ..., M$ denotes the probability that $Y = y \in \mathbb{N}^M$ when the message m is sent. The test strategy separates the observation space $\mathbb{N}^M$ into $M + 1$ domains of decision defined by

$$D_m = \{y \in \mathbb{N}^M | p(y|m) > p(y|l) \ \forall l \neq m\} \text{ and } D_c = \mathbb{N}^M - \bigcup_{m=1}^{M} D_m$$

$$\text{(II.1.1)}$$

The test works as follows:

  – for $y \in D_m$, m is selected

  – for $y \in D_c$, any message is selected with equal probability

The probability of error $\mathbf{P}_e$ can be bounded by computing the probability that the observation falls in any of the M domains D_m, so that

$$\mathbf{P}_e < 1 - \sum_{i=1}^{M} \left(\frac{1}{M} \sum_{y \in D_m} p(y|m) \right) \qquad \text{(II.1.2)}$$

The hypotheses being independent, we have:

$$p(y|m) = p_1(n_m) \prod_{\substack{i=1 \\ i \neq m}} p_0(n_i)$$

Moreover, the domains can now be redefined using the likelihood ratio $\Lambda(n_j) = \frac{p_1(n_j)}{p_0(n_j)}$. Hence $D_m = \{y \in \mathbb{N}^M | \Lambda(n_m) > \Lambda(n_i) \ \forall i \neq m\}$. Because $\Lambda(n_j)$ is a monotonic increasing function of n_j, we set $D_m = \{y \in \mathbb{N}^M | n_m > n_i \ \forall i \neq m\}$.

Therefore

$$p(y \in D_m | m) = \sum_{y \in D_m} p(y|m) = \sum_{n_m=0} \sum_{\substack{n_i=0 \\ i \neq m}} \cdots \sum p(y|m)$$

(II.1.3)

$$= \sum_{n=1} p_1(n) \left[\sum_{l=0}^{n-1} p_0(l) \right]^{M-1}$$

Finally,

$$P_e \leq \sum_{n=0}^{\infty} p_1(n) \left(1 - \left[1 - \sum_{l=n}^{\infty} p_0(l) \right]^{M-1} \right)$$

(II.1.4)

This test consists of evaluating $\Lambda(k)$ for the particular measured photon number and then comparing it with the λ, the threshold of the test.

II.1.1b Binary Hypothesis Testing

In a binary detection problem, the decision consists of a choice between two hypotheses, H_0 and H_1 (signal absent and signal present respectively) with the associated probability distributions p_0 and p_1 and a priori probability ζ_0 and ζ_1. In this case, the hypothesis H_1 is chosen whenever the observed photon number in the time interval $[0, T] : N(T) = n$, exceeds a certain integer decision level λ, and the hypothesis H_0 is chosen when $N(T) = n < \lambda$. For $N(T) = n = \lambda$, H_1 is again selected with a probability $f (0 \leq f < 1)$.

The threshold is chosen based on some criterion: if Q_d, the probability of detection, is maximized subject to a constraint on the probability of a false alarm (Neyman–Pearson criterion), the probability of detection is given by

$$Q_d = \Pr\{H_1|H_1\} = fp_1(\lambda) + \sum_{k=\lambda+1}^{\infty} p_1(k) \qquad (\text{II}.1.5)$$

where

$$Q_0 = \Pr\{H_1|H_0\} = fp_0(\lambda) + \sum_{k=\lambda+1}^{\infty} p_0(k) \qquad (\text{II}.1.6)$$

is the probability of false alarm which is the probability of choosing "signal present" (hypothesis H_1) when "no signal" (hypothesis H_0) is true. The required values of λ and f are obtained as indicated with a *preassigned* value of Q_0 and *given* $p_0(k)$ and $p_1(k)$. The average probability of error is then given by $P_{er} = \frac{1}{2}(1 + Q_0 + Q_d)$. As proved [6], for a Neyman–Pearson test the values of Q_0 and Q_d completely specify the performance: the curves Q_d versus Q_0 are called the Receiver Operating Characteristic "ROC".

In the next section, we will briefly present the method of computing the ROC curves for the case of multiple hypotheses.

II.1.1c Multiple Alternative Hypotheses

Suppose the signal to be detected belongs to a set of cardinality M, depending, for example, on a parameter $\Theta = \{\theta\}_{l=1}^{M-1}$. We assume that the detection process remains basically binary H_0 (signal absent) and H_1 one of the H_l ; $l = 1, ..., M - 1$ (signal present). To simplify the presentation, we assume that all the hypotheses H_l are equally likely. It is convenient to define the test vector $\Phi(X) = (\phi_0, ..., \phi_l, ...\phi_M)$ where

$$\phi_l(X) \overset{\text{def}}{=} \text{Prob [decide } H_l|X = x]$$

of the single observation $X = x \in \mathcal{X}$ so that $0 \leq \phi_l(X) \leq 1$; $\sum_{l=0}^{M} \phi_l(X) = 1$.

With the Neyman–Pearson criterion used, we fix the probability of false alarm

$$Q_0 = \int_{\mathcal{X}} \left(1 - \phi_0(X = x)\right) p_0(x) dx$$

at some level α and seek the test function which maximizes the probability of correct detection

$$Q_d = \int_{\mathcal{X}} \left(1 - \phi_0(X = x)\right) p_l(x) dx$$

Here, the generalized likelihood ratio $\Lambda_\ell(X) \overset{\text{def}}{=} \dfrac{p_l(X)}{p_0(X)}$ is combined with a randomization to read

$$1 - \phi_0(X) = \begin{cases} 1, & \max_l \Lambda_l(X) > \lambda \\ f_l, & \Lambda_l(X) = \lambda \\ 0, & \Lambda_l(X) < \lambda \end{cases} \qquad (\text{II.1.7})$$

where λ and f_l are chosen to achieve a preassigned value of $Q_0 = \alpha$. As previously, for photon counting purposes, the ROC curves are such that

$$Q_d = \Pr\{H_l|H_l\} = f_l p_l(\lambda) + \sum_{k=\lambda+1} p_l(k) \qquad (\text{II.1.8})$$

$$Q_0 = \Pr\{H_l|H_0\} = f_l p_0(\lambda) + \sum_{k=\lambda+1} p_0(k) \qquad (\text{II.1.9})$$

$$0 \leq f_l < 1 ; \quad \sum_{l=0}^{M} f_l = 1 \qquad (\text{II.1.10})$$

Before closing this brief introduction to the statistical detection theory, we wish to mention another criterion which requires maximizing the signal to noise ratio. In fact, let the random variable X be a sufficient statistic of the data and suppose we want to decide between two hypotheses that depend on the real signal parameter μ, $H_1 : p_1(x, \mu)$, $H_0 : p_0(x, 0)$. We may define the signal to noise ratio as

$$\Delta = \frac{(\mathrm{E}[X|H_1] - \mathrm{E}[X|H_0])^2}{\mathrm{E}[X^2|H_0] - (\mathrm{E}[X|H_0])^2} \tag{II.1.12}$$

where $\mathrm{E}[X^k|H_i] = \int x^k p_i(x)dx$, $i = 0, 1$. A processor that maximises Δ for $\mu \to 0$ is called a *threshold* detector.

II.1.2 Information

For a discrete memoryless channel without cost constraint which is the conventional modelisation of an optical channel, a few fundamental results are required to derive the expressions of some information criteria: capacity, cut–off rate and rate distortion function.
We begin with briefly recalling some elements of the classical theory and then extend some results to the quantum case in the next section.

II.1.2a Entropy and Algorithmic Complexity

The axiomatic approach, rather mathematical, to justify the concept of entropy and derive its properties, will not be mentioned here.
The statistical thermodynamics point of view is first adopted. The probability p_m for a system to be in state $|m\rangle$ characterized by the energy E_m, yields the entropy $S = -k_B \sum_{m \in \mathcal{M}} p_m \log p_m$, with

$\sum_m p_m = 1, 0 \leq p_m \leq 1, p_m = p_m^*$ and where k_B is the Boltzmann's constant.

To treat an example, let be given an ensemble of N noninteracting particles (e.g. spins $\{\sigma_i\}$ such that $\langle \sigma_i \rangle = \sigma$). Let us consider that $q_+ = \frac{N+n}{2}$ particles (resp. $q_- = \frac{N-n}{2}$) are oriented up $(+1)$ (resp. down (-1)). The total number of the possible sequences of such order obeys the Bernouilli law $W(n) = \binom{N}{q_+}(\frac{1}{2})^{q_+} + (\frac{1}{2})^{q_-}$, $(q_+ + q_- = 1)$. Under the action of external field of intensity 1, each particle (i, σ_i), has a probability $p_i(+1)$ (resp. $p_i(-1)$) to be oriented so that the total energy $U = \sum_{i=1}^{N} \sigma_i = N(p_i(+1) - p_i(-1))$, from which we deduce $p_i(\pm 1) = \frac{1}{2} \pm \frac{U}{2N}$.

To calculate $W = \frac{1}{p(U)}$, it is useful to utilize the partition function $Z = \sum_{\sigma_i = \pm 1} e^{-\beta U} = \prod_{i=1}^{N} \sum_{\sigma_i = \pm 1} e^{-\beta \sigma_i} = (e^{-\beta} + e^{\beta})^N = \sum_{q=0}^{N} \binom{N}{q} e^{-\beta(2q-N)}$, where $\beta = \frac{1}{k_B T}$, T is the temperature. Assuming that all states of energie $U \pm \Delta U$ are of equal probability, we may deduce that $W \sim Z \sim \binom{N}{q}$ with $q_\pm \sim \frac{U \pm N}{2} = N p_i(\pm)$.

Using the Stirling formula,

$$\log W = \log N! - \log q! - \log(N-q)! = N \log N - q_+ \log q_+ - q_- \log q_-$$
$$= N \left(\log N - p_i(+1) \log(N p_i(+1)) - p_i(-1) \log(N p_i(-1)) \right)$$
$$= -N \sum_{\sigma_i = \pm} p(\sigma_i) \log p(\sigma_i) = \frac{N}{k_B} S$$

which is of the same form as the previous definition.

It should be stressed that despite the fact that there is a close correspondence between thermodynamical and statistical quantities, their relationship with the information entropy is not clearly established.

We now introduce the concept of entropy as a measure of uncertainty of a random variable $X : (x, p(x))$ as expected value of the random variable $q(x) = \log \frac{1}{p(x)}$

$$S(X) = \sum_{x \in \mathcal{X}} p(x) q(x) = - \sum_{x \in \mathcal{X}} p(x) \log p(x)$$

With the aid of the joint probability $p(x, y)$ and the conditional probability $p(y|x)$ of a pair of random variables X, Y, the above entropy is readily extended to the joint entropy

$$S(X, Y) = -\sum_{x,y} p(x, y) \log p(x, y)$$

and to the conditional entropy $S(Y|X) = -\sum_{x,y} p(x, y) \log p(y|x)$. It is therefore simple to introduce the mutual information

$$I(X; Y) = S(Y) - S(Y|X)$$

which is a quantitative measure of information one random variable contains about the other on one hand, and the relative entropy that characterizes the distance of two probability distributions $S(p\|q) = \sum_x p(x) \log \frac{p(x)}{q(x)}$ on the other hand.

Another extension concerns a continuous random variable for which the *differential entropy* can be defined as

$$S_f(X) = -\int_{\mathcal{X}} f(x) \log f(x) dx, \quad \int_{\mathcal{X}} f(x) dx = 1$$

This function has, however, a limited interest because it can diverge or can be negative. Nevertheless, it is often justified to simplify the calculations as shown in the next conventional example.

For a one dimension time–dependent system $\{X : x, p(x, t)\}$ for which the following constraints $\int_{\mathcal{X}} x^k p(x, t) dx = m_k, k = 0, 1, 2$, are given such that $m_0 = 1, m_1(t), m_2$, it is readily shown that the maximization of $S_p(t) = -\int_{\mathcal{X}} p(x, t) \log p(x, t) dx$, the differential entropy, using the method of Lagrange multipliers yields the Gaussian distribution $p(x, t) = \mathcal{N}(m_1, \sigma)$ where the variance is $\sigma^2(t) = m_2 - m_1^2(t)$.

To do so, we may interpret $m_1(t) = m_1(0)e^{-\lambda t}$ as the solution of the first order differential equation $\frac{d}{dt}m_1(t) + \lambda m_1(t) = 0$ which is equivalent to $\frac{\partial p(x,t)}{\partial t} = \frac{\partial p(x,t)}{\partial m_1}\frac{dm_1}{dt} = -\left(\frac{\partial p(x,t)}{\partial x} + m_1\frac{\partial^2 p(x,t)}{\partial x^2}\right)\frac{dm_1}{dt} = \lambda m_1 \frac{\partial p(x,t)}{\partial x} + \lambda m_1^2 \frac{\partial^2 p(x,t)}{\partial x^2}$ and leads to $\frac{\partial p(x,t)}{\partial t} = \frac{\partial}{\partial x}\left(\lambda x\ p(x,t)\right) + \frac{\partial^2}{\partial x^2}\left(\lambda m_2\ p(x,t)\right)$, the well–known Fokker–Planck equation (FPE). To prove the result, it is useful to notice that, for a normal distribution $p(x,t) = \mathcal{N}(m_1,\sigma)$, $p(x,t) = -(x - m_1)\frac{\partial p(x,t)}{\partial x} - \sigma^2\frac{\partial^2 p(x,t)}{\partial x^2}$, and $\frac{\partial p(x,t)}{\partial m_1} = -\frac{\partial p(x,t)}{\partial x} - m_1\frac{\partial^2 p(x,t)}{\partial x^2}$.

Since the difficulties due to the definition of differential entropy disappear when studying relative entropy or mutual information, it is therefore, preferable to deal with relative entropy and not to define differential entropy at all. In that case, a general rigorous theory can be developed from the point of view of the measure theory [29].

As it is just seen, the definition of entropy requires the knowledge of the probability distribution of the observable. However, the concept of probability, that is essential in quantum mechanics, is not necessary in classical physics which is a deterministic approach. That the reason why, certain authors (e.g. Zurek [27]) reinterpreted the usual concept of entropy as the *algorithmic complexity* to specify the state of the system under observation, without utilizing any notion of probability.

Let us briefly introduce the concept of algorithmic complexity in order to see the sort of link it is made with the entropy. Let $\mathcal{C}(p)$ be the output of the computer $\mathcal{C}$ processing the numerical program p. Now, let $l(p)$ denote the length (integer) of a finite length string $X \stackrel{\text{def}}{=} x$. The algorithmic complexity that depends on $\mathcal{C}$ via a constant $K_{\mathcal{C}}(X) = \min_{\mathcal{C}(p)=X} l(p)$ is the minimal length of a program to obtain X. As a first approximation, the complexity can be taken equal to the length of the string.

Example:

Let us order a sequence of N positive numbers $a(k) \geq 0$, $k = 1, ..., N$, each number requiring, e.g. $\log_2 a_m$ to be coded. The output string $\{a(k)\}$ is ordered so that the numbers are in increasing values. Therefore $K(X) \sim \lceil \log_2 N \rceil + N \lceil \log_2 a_m \rceil$ ($\lceil x \rceil$ is the smallest integer $\geq x$).

Remarks:

(i) From the Kraft inequality $\sum_i 2^{-l_i} \leq 1$, and the convexity property of the log function $\left(\log \sum_i p_i \left(\frac{2^{-l_i}}{p_i} \right) \geq \sum_i p_i \log \frac{2^{-l_i}}{p_i} \right)$, (log taken in base 2), $\sum_i p_i \left(\frac{2^{-l_i}}{p_i} \right) \leq 1$, $\log \sum_i p_i \left(\frac{2^{-l_i}}{p_i} \right) \leq 0$, then

$$\sum_i p_i \log \frac{2^{-l_i}}{p_i} \leq 0 \;\; \rightarrow \;\; \mathrm{E}[l_i] \geq S(p_i)$$

(ii) The probability of a string x [32, p. 160] can then be introduced :

$\Pr(\mathcal{C}(p) = x) = \sum_{p:\mathcal{C}(p)=x} 2^{-l(p)}$.

The relationship between complexity and entropy is established for i.i.d. random variables $\{X_i\}_{i=1}^{N}$: $x_i, p(x_i)$ for which we denote $X : x, p(x) = \prod_i^N p(x_i)$. The theorem [32, p.154] states that

$$\lim_{N \to \infty} \mathrm{E}\left[K(X|N)\right] - S(X)$$

where $S(X) = N S(X_i)$ is the entropy of the stochastic process X. For a given $l(X) = x$, the complexity is written as a conditional complexity $K_{\mathcal{C}}(X|x)$ from which it will be derived $I(X;Y) = K(X) - K(Y|X)$ that is interpreted as a quantitative measure of information about Y contained in X.

Notice, however the equality $I(X;Y) = I(Y;X)$ cannot be expected to hold exactly.

A bibliography concerning the questions of entropy and complexity and their connections and relationships with other fields can be found in [32, and references therein]. Notice, however, that it seems difficult to formulate the notion of algorithmic complexity for quantum systems [30].

II.1.2b Channel Capacity

Let the discrete channel input sequence be $X = \{x_i = i\}_{i=1}^{M}$ given with the probability $q(x_i)$ and the output sequence $Y = \{y_j = j\}_{j=1}^{N}$. The statistical description of the channel is determined by the Π matrix $N \times M$ of elements being the conditional probability distribution $p(y_j|x_i)$. The mutual information between X and Y can be written in the usual form

$$I(X,Y) \overset{\text{def}}{=} S(X) - S(X|Y)$$

$$= \sum_{i=1}^{M} \sum_{j=1}^{N} q(x_i)p(y_j|x_i) \log \frac{p(y_j|x_i)}{\sum_k q(x_k)p(y_j|x_k)} \qquad \text{(II.1.13)}$$

The channel capacity is the maximum under the constraints Q of the mutual information rate (II.1.13)

$$\begin{cases} C = \max_{Q} \; I(X;Y) \\ Q : \sum_{i=1}^{M} q(x_i) = 1 \end{cases} \qquad \text{(II.1.14}a,b)$$

Theorems on convexity of the mutual information, are demonstrated in various textbooks (see [3,29]). Recall that $I(X;Y)$ is convex $\cap$ with respect to $q(x_i)$, $\forall i$ and convex $\cup$ with respect to $p(y_j|x_i)$, $\forall i,j$. Let us show here a simplified calculation [8]. We use a Lagrange method for the maximisation of (II.1.14a,b) to seek the probabilities $\{q(x_i)\}$ such that

$\forall i, \dfrac{\partial \Lambda}{\partial q(x_i)} = 0$ where $\Lambda = S(X) - S(X|Y) + \lambda \displaystyle\sum_{i=1}^{M} q(x_i) - 1$. After simple

algebra, it is shown that the system (II.1.14a,b) reduces to

$$\sum_{j=1}^{N} p(y_j|x_i) \log \frac{p(y_j|x_i)}{p(y_j)} + \lambda - 1 \equiv I(x_i; Y) - C = 0 \qquad \text{(II.1.14c)}$$

where $C \overset{\text{def}}{=} 1 - \lambda$ is called the channel capacity. C is exactly the channel capacity because for $q(x_i)$ satisfying (II.1.14c) we do have that
$\sum_i q(x_i) I(x_i; Y) = C \overset{\text{def}}{=} \max I(X; Y)$. To determine $\{q(x_i)\}$, suppose
that $M = N$ and assume that the matrix $\mathbf{P}$ of the transition probabili-
ties $p(y_j|x_i)$ is not singular. Therefore

$$\forall i = 1, ..., M \qquad \sum_{j=1}^{M} \Big[C + \log p(y_j) \Big] p(y_j|x_i) = -S(Y|x_i)$$

$$\forall j = 1, ..., M \qquad \begin{cases} p(y_j) = \displaystyle\sum_{i=1}^{M} q(x_i) p(y_j|x_i) \\ \displaystyle\sum_{j=1}^{M} p(y_j) = 1 \end{cases}$$

$$\text{(II.1.14}d, e)$$

Inverting (II.1.14d,e) yields

$$C = \log \sum_{j=1}^{M} \exp \left(-\sum_{i=1}^{n} \pi_{ij} S(Y|x_i) \right)$$

$$q(x_i) = \exp(-C) \sum_{j=1}^{M} \pi_{ij} \exp \left(-\sum_{k=1}^{n} \pi_{jk} S(Y|x_k) \right) \qquad \text{(II.1.15}a, b, c)$$

$$p(y_j) = \exp(-C) \exp \left(-\sum_{i=1}^{n} \pi_{ji} S(Y|x_i) \right)$$

where π_{ij} are elements of $\mathbf{P}^{-1}$. However, it may happen that $q(x_i) < 0$.
To avoid such case and guarantee that $q(x_i) \geq 0$, the standard Kuhn–
Tucker conditions [32, p.462] can be used to characterize the maximum.
They are equivalent to $\frac{\partial}{\partial q(x_i)} \Lambda|_{q(x_i)} = 0$ for $q(x_i) > 0$ and $\frac{\partial}{\partial q(x_i)} \Lambda|_{q(x_i)} \geq$
0 for $q(x_i) = 0$, where Λ is the Lagrangian function.

Here

$$Q' : \sum_{i=1}^{M} q(x_i) - 1 \le 0, \quad -q(x_i) \le 0 \qquad \text{(II.1.15d)}$$

must be substituted into (II.1.14b) and numerical calculation are very simple. The algorithm of computation is summarized as follows.

Let us be given the vector $\overline{q} = \overline{q}(x) : x = \{x_i\} \in \mathcal{X}, \sum_x \overline{q}(x) = 1$ and the matrix $\overline{Q} = \overline{Q}(x|y) : y = \{y_j\} \in \mathcal{Y}, \overline{Q}(x|y) \ge 0, \sum_x \overline{Q}(x|y) = 1 \ \forall y.$
Define $F(\overline{q}, \overline{Q}) = \sum_{x,y} p(y|x)\overline{q}(x) \log \frac{\overline{Q}(x|y)}{\overline{q}(x)}$, we first get $\frac{\partial F(\overline{q}, \overline{Q})}{\partial \overline{q}(x_i)} = \lambda_i =$
$\sum_{y_j} p(y_j|x_i) \log \overline{Q}(x_i|y_j) - \sum_{y_j} p(y_j|x_i)\big(\log \overline{q}(x_i) + 1\big)$
$= \sum_{y_j} p(y_j|x_i) \log \overline{Q}(x_i|y_j) - \big(\log \overline{q}(x_i) + 1\big)$, λ_i being derived from $\sum_x \overline{q}(x) = 1$ to yield $\overline{q}(x_i)$ given by $R2$ and then

$$
\begin{cases}
Q^{(k)}(x_i|y_j) = \dfrac{p(y_j|x_i)q^{(k)}(x_i)}{\displaystyle\sum_{x_i'} p(y_j|x_i')q^{(k)}(x_i')} \\[4ex]
q^{(k+1)}(x_i) = \dfrac{\exp\left(\displaystyle\sum_{y_j} p(y_j|x_i) \log Q^{(k)}(x_i|y_j)\right)}{\displaystyle\sum_{x_i'} \exp\left(\displaystyle\sum_{y_j} p(y_j|x_i') \log Q^{(k)}(x_i'|y_j)\right)}
\end{cases}
\qquad (R1, R2)
$$

Starting from arbitrary $q^{(0)}$ (e.g. $1/M$), the steps of the computation follow

$$q^{(0)}(x) \xrightarrow{(R1)} Q^{(0)}(x|y) \xrightarrow{(R2)} q^{(1)}(x) \to C(1) \equiv F(q^{(1)}, Q^{(0)})$$

so that $C(k+1) = F(q^{(k+1)}, Q^{(k)})$. For ϵ such that $|C(k+1) - C(k)| < \epsilon$, the convergence can be proved $\lim_{k \to \infty} |C - C(k)| = 0$.

For several types of channel, efficient numerical algorithms are available and can be easily implemented [32, p. 364].

As for the above discrete case, a simplified version of the capacity calculation can be presented [32, p. 241].

Let us assume that a channel superimposes an independent white Gaussian noise Z, $g(z) \in \mathcal{N}(0, \sigma_Z^2)$ to a continuous source X of a probability density $f(x), x \in \mathcal{X}$ and a differential entropy $S_f(X)$.

The channel output is $Y = X + Z$ and we have

$$h(y) = f(x)g(z), \quad \langle Y^2 \rangle = \sigma_Z^2 + \langle X^2 \rangle = N_0 + P, \quad S_g(Z) = \frac{1}{2}\log 2\pi e \sigma_Z^2$$
$$\text{(II.1.16)}$$

The channel capacity

$$C \overset{\text{def}}{=} \max_{(\mathcal{C})} I(X;Y) \quad \text{where } (\mathcal{C}) = \begin{cases} f(x) \geq 0, \quad \displaystyle\int_x f(x)dx = 1 \\ \displaystyle\int_x x^2 f(x)dx \leq P \end{cases}$$
$$\text{(II.1.17a, b)}$$

We now use the fundamental lemma which yields $S_f(Y) \leq S_g(Y)$, for given P, N_0, to obtain

$$\begin{aligned} I(X;Y) &= S(Y) - S(Y|X) = S(Y) - S(X + Z|X) \\ &= S(Y) - S(Z|X) = S(Y) - S(Z) \\ &\leq \frac{1}{2}\log 2\pi e(P + N_0) - \frac{1}{2}\log 2\pi e N_0 = \frac{1}{2}\log\left(1 + \frac{P}{N_0}\right) \end{aligned}$$

which finally yields

$$C = \frac{1}{2}\log\left(1 + \frac{P}{N_0}\right) \qquad \text{(II.1.18a)}$$

The capacity is attained for $X \in \mathcal{N}(0, P)$. When the constraint on the bandwidth $2W$ of a white noise is included,

$$C = W\log\left(1 + \frac{P}{N_0 W}\right) \qquad \text{(II.1.18b)}$$

$W \to \infty, C \to \frac{P}{N_0}$

$W \to 0, C \sim W\log\frac{P}{N_0 W} = W\log\frac{P}{N_0} - W\log W \to 0$

For a colored Gaussian noise [7], (II.1.18a) becomes

$$C = \frac{1}{2}\int_{-\infty}^{\infty} \log\left(1 + \frac{\gamma_g(\nu)}{\gamma_B(\nu)}\right)d\nu \qquad \text{(II.1.18c)}$$

where $P = \displaystyle\int_{-\infty}^{\infty} \gamma_S(\nu)\,d\nu$ and $N = \displaystyle\int_{-\infty}^{\infty} \gamma_B(\nu)\,d\nu$.

To continue with the example, let us specify the type of the input source X as M orthogonal signals. The decision rule which is optimum in the sense that the probability of error is mimimum defines the domains of decision $\{\Lambda_m\}$.

For equal *a priori* probabilities

$$\Lambda_m = \Big\{ \mathbf{y} : \ \log p(\mathbf{y}|\mathbf{x}_m) > \log p(\mathbf{y}|\mathbf{x}_{m'}) \Big\}$$
$$= \Big\{ \sum_n 2(x_{mn} - x_{m'n})y_n > \sum_n (x_{mn}^2 - x_{m'n}^2) \Big\}$$

for $\forall m' \neq m$. To calculate the probability of error, we first study the
case $M = 2$. We notice that $P_c = p(y \in \Lambda_1|H_1) = \int \int_{\Lambda_1} f(y_1, y_2) dy_1 \, dy_2$.
Thus

$$P_c = \frac{1}{2\pi N_0} \int_{-\infty}^{\infty} dy_1 \exp\left(-\frac{(y_1 - \sqrt{E})^2}{2N_0}\right) \int_{-\infty}^{y_1} dy_2 \exp\left(-\frac{y_2^2}{2N_0}\right)$$

$$= \frac{1}{\sqrt{2\pi}} \int_{-\infty}^{\infty} d\theta_1 \exp\left(-\frac{\theta_1^2}{2}\right) \frac{1}{\sqrt{2\pi}} \int_{-\infty}^{\sqrt{\frac{E}{N_0}}+\theta_1} d\theta_2 \exp\left(-\frac{\theta_2^2}{2}\right)$$

$$= \frac{1}{\sqrt{2\pi}} \int_{-\infty}^{\infty} \exp\left(-\frac{\theta^2}{2}\right) \left(1 - Q\left(\sqrt{\frac{E}{N_0}} + \theta\right)\right) d\theta$$

For arbitrary M, using the M–ary property: the signal is transmitted
only if $x_i \rightarrow y_i$, such that given x_i, we have $y_i > y_j, \forall j \neq i$ and

$$\Pr(y_1|x_i) = ... = \Pr p(y_j|x_i) = ... = \Pr p(y_M|x_i)$$

so that

$$q(y_i) = \Pr(y_j|x_i) = \int_{-\infty}^{y_i} \mathcal{N}(0, \frac{N_0}{2}) dy_j$$

Thus $M - 1$ "equal" components contribute to the total probability

$$P_c = \int_{-\infty}^{\infty} (q(y_i))^{M-1} p(y_i|H_i) dy_i$$

Noticing that for $\phi_M(t)$, $M = 1, 2, ...$, and for any $g(t)$ so that $|\phi_M(t)| \leq g(t) - \forall t, M$ and $\int_{-\infty}^{\infty} g(t)dt < \infty$ we have $\lim_{M \to \infty} \int_{-\infty}^{\infty} F(M, \theta) d\theta = \int_{-\infty}^{\infty} \lim_{M \to \infty} F(M, \theta) d\theta$. Therefore [5]

$$P_e = 1 - P_c = 1 - \int_{-\infty}^{\infty} p(y|H_1) \left(\int_{-\infty}^{y} p(x|H_0) dx \right)^{M-1} dy$$

$$\leq (M - 1) \frac{1}{\sqrt{2\pi}} \int_{\sqrt{\frac{E}{N_0}}}^{\infty} e^{-\frac{\theta^2}{2}} d\theta \leq \frac{M - 1}{\sqrt{\frac{E}{N_0}} \sqrt{2\pi}} e^{-\frac{E}{2N_0}}$$

To study the asymptotic case $M \to \infty$, we set the following notations $M = e^{RT}, R = \frac{1}{T} \log M, E = \frac{P}{R} \log M$ with $E \leq PT$,

$$P_e(M) \sim 1 - e^{-c\mathcal{E}(M)}$$

$$\mathcal{E}(M) = M^{1-\frac{C}{R}} \begin{cases} R > C : \dfrac{C}{R} < 1 & \mathcal{E}(M) \sim M \to \infty & P_e \to 1 \\[2mm] R < C : \dfrac{C}{R} > 1 & \mathcal{E}(M) = \dfrac{1}{M} \to 0 & P_e \to 0 \end{cases}$$

$$\text{(II.1.19)}$$

Before closing the section on capacity criterion, let us just express the fundamental theorem channel coding:

Given $M = e^{NR} \to \infty$ equally likely signals, a constant C exists, called the Gaussian channel capacity whose properties are such that

  – $R > C$, the P_e probability of error $\to 1$

  – $R < C$, the P_e probability of error achieved with optimum receivers $\to 0$

Rigorous demonstration of the capacity theorems (direct and converse) for Gaussian channel can be found in [3,5].

Remark: Notice that the shape of the signals is not important. We here again mention that the above discussion is only a first approach.

II.1.2c Cut–Off Rate

To deal with the cut-off rate, we first define some notations and simplify the presentation of some results.

For each pair x_1, x_2 taken from a finite set, the source alphabet A_X, supposed to be i.i.d. with common distribution function $\text{Prob}(X = x) = Q(x)$ called the source statistics, we introduce

$$J(x_1, x_2) = \sum_{y \in A_Y} \sqrt{p(y|x_1)p(y|x_2)}$$

where the finite set A_Y is called the destination alphabet. Let

$$J_0 = \min_{Q(x)} \left\{ E[J(x_1, x_2)] \right\}$$

Consider a code of length N containing two codewords $X_1 = \{x_{1i}\}_{i=1}^N$, $X_2 = \{x_{2i}\}_{i=1}^N$. Assume that given a received N-uple Y, the decoder output is

- X_1 if $p(Y|X_1) > p(Y|X_2)$
- X_2 if $p(Y|X_2) > p(Y|X_1)$.

Let $Y_1 = \{Y : p(Y|X_1) > p(Y|X_2)\}$ and $Y_2 = \{Y : p(Y|X_2) > p(Y|X_1)\}$. If $P_e^{(i)}$ denotes the probability of error of decoding, given that X_i is sent, we can show that

$$P_e^{(i)} \le \sum_{y \in Y_i} p(y|X_i) \le \sum_{y \in A_Y^n} \sqrt{p(Y|X_1)p(Y|X_2)} = \prod_{k=1}^N J(x_{1k}, x_{2k})$$

$$(\text{II.1.20})$$

The generalisation to M codewords is $P_e^{(i)} \le \sum_{\substack{j=1 \\ j \ne i}} \prod_{k=1}^N J(x_{ik}, x_{jk})$.

Furthermore by averaging over all codes, we obtain $E\left[P_e^{(i)}\right] \le M \exp(-N R_0)$. Finally, the cut-off rate in nats units, is given by

$$R_0 = -\log(J_0) \qquad (\text{II.1.21})$$

With regard to this cut-off rate function, the important result proved by McEliece [18] can be summarized as follows

For a DMC and for any rate less than the cut-off rate $R < R_0$, a code denoted $\{x_1, x_2, ..., x_M\}$ exists with at least $M = \exp(R_N)$

codewords of length N and an appropriate decoding rule, such that if $P_e = \frac{1}{M} \sum_{i=1}^{M} P_e^{(i)}$ *denotes the average decoding error probability,* *then*

$$P_e < \exp\left(-N\left(R_0 - R\right)\right) \tag{II.1.22}$$

This theorem:

- is stronger than the channel coding in that it gives an explicit estimate of how small P_e can be made as a function of n.
- is weaker in that $\forall R, \; R < C$.

Therefore for rates $R_0 < R < C$ it does not imply that $P_e \to 0$ is at all possible.

II.1.2d Rate Distortion

The purpose of this section is to briefly summarize some elements of the theory of the rate distortion functions for discrete memoryless sources. We begin with introducing notations and simplifying the presentation of a few results.

Consider a *discrete, noiseless, memoryless channel* that maps the source $\left(X = \{j\}_1^M, \text{Prob}\, (X = j) = p_j\right)$ into the destination $\left(\widehat{X} = \{k\}_1^N, \text{Prob}\, (\widehat{X} = k) = r_k\right)$ by the transition probability $q(k|j), \; \forall j, \sum_{k=1}^{N} q(k|j) = 1$.

Suppose that for each pair (j, k), a nonnegative number Δ_{jk} exists which measures the degradation (*error distortion*) that the symbol j is reproduced by the symbol k and does not depend on any other terms. Its average is

$$\langle \Delta_{jk} \rangle = \sum_{j,k} \Delta_{jk} p_j q(k|j) = d(q) \tag{II.1.23a}$$

The mutual information is as usual

$$I(X;\widehat{X}) = I(q) = \sum_{j,k} p_j q(k|j) \log \frac{q(k|j)}{r_k} \qquad (\text{II.1.23}b)$$

Now, given p_j, Δ_{jk} and D, the $R(D)$ rate–distortion function is simply

$$\begin{cases} R(D) = \min_{q \subset Q_D} I(q) \\ Q_D = q(k|j) : d(q) \le D \end{cases} \qquad (\text{II.1.24})$$

the minimum does exist because the function $I(q)$ is convex $\cup$ with respect to the $q(k|j)$. Thus, the new source $\widehat{X}$ will reproduce some statistical properties of the initial source X with lower entropy. The $R(D)$ is the minimum bit rate required to encode X subject that the distorsion is maintained arbitrary close the level D.

Assuming that $M = N$ and using the well–known method of the Lagrange multipliers and the conditions for making sure that all probabilities are positive and normalised [32, p. 462], the appropriate solution is determined such that

$$\begin{cases} s < 0 \\ c_k = \sum_j \lambda_j p_j \exp(s\Delta_{jk}) \le 1, \forall k \\ q(k|j) = \lambda_j r_k \exp(s\Delta_{jk}) \\[2mm] D = \sum_{j,k} \Delta_{jk} p_j q(k|j) \\ D_m = \sum_j p_j \min_k (\Delta_{jk}), \quad D_M = \min_k \sum_j p_j \Delta_{jk} \\ R(D) = sD + \sum_j p_j \log \lambda_j \quad D \in [D_m, D_M] \end{cases} \qquad (\text{II.1.25}a)$$

where D_m and D_M are the minimum and the maximum values of D respectively. A more general expression of $R(D)$ involves real matrices formulation.

Denote Π the matrix of elements $\{\pi_{jk} = p_j q(k|j)\}$ obtained from (II.1.25a), A the matrix of elements $\{a_{jk} = e^{s\Delta_{jk}}\}$, $\{\omega_{jk}\}$ elements of the inverse matrix A^{-1}, $\overrightarrow{p}$ the vector of elements $\{p_j\}$, Δ the matrix of elements $\{\Delta_{jk}\}$, then

$$R(D) = S(\overrightarrow{p}) + sD(s) + \sum_j p_j \log \mu_j \qquad (\text{II.1.25}b)$$

where $S(\overrightarrow{p}) = -\sum_j p_j \log p_j, D = \text{Tr } (\Pi\Delta^+), \mu_j = \lambda_j p_j = \sum_k \omega_{jk}$.

For the distorsion measure selected here $\Delta_{jk} = \Delta_{|j-k|} = \Delta_l$ which includes the probability of error criterion $\Delta_{jk}^e = 1 - \delta_{jk}$, Eq. (II.1.25b) can interestingly lower bounded.

To find the bound, we use the fact that for some $|s| > |s^*| \gg 1$, we have $a_{jk} = a_l$ such that $a_0 = 1 \geq a_1 \geq ... \geq a_M$. The calculation of the inverse matrix A^{-1} becomes easy yielding $\{\mu_j\} \sim \left\{(1-(M-1)a_1)^2\right\} \forall j, (M \gg 1)$. The average value of the distorsion will then be given by

$$D^*(s) = \sum_{j,k,l} \omega_{kl} p_l \frac{\mu_j}{\mu_l} \Delta_{jk} e^{s\Delta_{jk}}$$

$$\sim (M-1)a_1 \sum_{k,l} \omega_{kl} p_l \sim \frac{(M-1)a_1}{1+(M-1)a_1} \qquad (\text{II.1.25}c)$$

where $a_1 = e^s$. It leads to a parametric expression of the rate distorsion

$$R(D^*) \sim S(\overrightarrow{p}) + sD^*(s) - \log\left(1+(M-1)a_1\right) \qquad (\text{II.1.25}d)$$

where $s(D) = \log \frac{D}{(M-1)(D-1)}$ and $|s^*| \sim s(D_M)$.

We finally summarize a few basic properties of $R(D)$ that will be utilized in the present context, among those proved by Berger [11]

- There is one and only one relative min of the mutual information with respect to the transition probability.
- $R(D)$ is a continuous, monotonic decreasing, convex $\cup$ function in the range $D \in [D_m, D_M]$.
- $R(0) = S(\overrightarrow{p})$, $R(D) = 0$ for $D \geq D_M$ ($\overrightarrow{p}$, the source a priori vector probability).

We close this section by treating two examples for Δ^e_{jk} and where the previous approximate calculations become exact.

First, for $M = N = 2$, $\Delta_{jk} = 1 - \delta_{jk}$, given $p_0 = p, p_1 = 1 - p$, it is proved that $r_0 = \frac{p-D}{1-2D}, r_1 = 1 - r_0$, $(0 \leq r_0 \leq p \leq \frac{1}{2})$. It is then easily derived that

$$\begin{cases} R(D) = S(p) - S(D) & 0 \leq D < p \\ R(D) = 0 & D \geq p \end{cases} \qquad (\text{II.1.26}a)$$

where $S(z) = -z \log z - (1-z) \log(1-z)$. Secondly, a generalisation to higher number of messages will show that the output vector probability $\overrightarrow{r}$ can be written formally as a linear function of the input vector probability $\overrightarrow{p}$

$$\overrightarrow{r} = \frac{1 + (M-1)a_1}{1 - a_1} \overrightarrow{p} - \frac{a_1}{1 - a_1} \overrightarrow{1} \qquad (\text{II.1.26}b)$$

which yields

$$\begin{cases} R(D) = S(\overrightarrow{p}) - D \log(M-1) - H(D) & \text{for } 0 \leq D \leq D_M \\ R(D) = 0 & \text{for } D \geq D_M \end{cases}$$

$$(\text{II.1.26}c)$$

where $D_M = \inf[1 - \max_j p_j, (M-1)\min_j p_j] \leq 1 - \dfrac{1}{M}$.

We refer the interested reader to the books [11,18] for more details.

II.2 Quantum Expressions

As has been done above for the classical formulation, the quantum expressions for the detection and informational entropy criteria will be briefly established.

Let us consider the system of communication [4,10] in which the input source of messages, the density operators $\{\widehat{\varPsi}_i\}_{i=0}^{M-1}$ emitted with the *a priori* probability and the measurement operators verify

$$
S = \begin{cases}
\forall i,\ \widehat{\varPsi}_i \geq 0,\, \mathrm{Tr}\,(\widehat{\varPsi}_i) = 1 \\[2mm]
\xi_i \geq 0,\ \displaystyle\sum_{i=0}^{M-1} \xi_i = 1 \\[2mm]
\forall j,\ \widehat{R}_j > 0,\ \displaystyle\sum_{j=0}^{N-1} \widehat{R}_j = \mathbb{1}
\end{cases}
\tag{II.2.1a}
$$

The detection channel is characterized by the transition probability

$$
p_{ij} = \mathrm{Tr}\,[\widehat{\varPsi}_i\widehat{R}_j],\ p_{ij} \geq 0,\ \sum_j p_{ij} = \zeta_i,\ \sum_{ij} p_{ij} = 1,\ N = M
\tag{II.2.1b}
$$

Hence the probability of error will be

$$
P_e = 1 - \sum_{j=0}^{M-1} \xi_i\, \mathrm{Tr}\,[\widehat{\varPsi}_i\widehat{R}_j]
\tag{II.2.1c}
$$

The problem of quantum optimal detection as formulated by Hel-strom and Holevo is to find the operators $\{\widehat{R}_j\}$ so that the P_e is minimum. Although considerable work and important results have been obtained over the last twenty years, a rigorous solution to this problem has not yet been given. However, for some interesting cases, exact results have been established. They are the discrimination between

- M linearly independent pure states

- M commuting density operators.

The detection problem is to make a choice, after a single measurement, between M hypotheses H_i each corresponding to $\{\widehat{\Psi}_i\}$. Let $\underline{C}$ be the specified Bayes cost matrix. The element C_{ij} is the cost of choosing hypothesis H_i when H_j is true $\forall i \neq j$, $C_{ij} > C_{jj}$. One has to minimize the average $\overline{C} = \sum_{ij} C_{ij}\xi_i \operatorname{Tr} \left[\widehat{\Psi}_i\widehat{R}_j\right]$. The $\{\widehat{R}_i\}$ making $C^* = \overline{C}_{min}$ is called optimum detection operator (o.d.o.). Let us denote $\widehat{\Upsilon} = \sum_j \widehat{R}_j\widehat{\Psi}_j$, $\widehat{W}_i = \sum_j \xi_j C_{ij}\widehat{\Psi}_j$ and $\widehat{\Gamma}_i = \widehat{W}_i - \widehat{\Upsilon}$. The necessary and sufficient conditions to verify, $\forall i,j$,   are

$$\begin{cases} \widehat{\Upsilon} = \sum_j \widehat{R}_j\widehat{\Psi}_j, & C^* = \operatorname{Tr}[\widehat{\Upsilon}] \\ \widehat{\Gamma}_i \geq 0, & \widehat{\Gamma}_i\widehat{R}_i = \widehat{R}_i\widehat{\Gamma}_i = 0 \end{cases} \qquad (\text{II}.2.2a, b)$$

II.2.1 Decision Among M Linearly Independent Pure States

We set $\widehat{\Psi}_i = |\psi_i\rangle\langle\psi_i|$. Taking the costs $C_{ij} = -\delta_{ij}$, the optimum p.o.m. is the set of projection–valued operators $\widehat{\Pi}_j = |r_j\rangle\langle r_j|, \sum_{j=1}^M \widehat{\Pi}_j = \mathbb{1}, \langle r_k|r_j\rangle = \delta_{kj}$, given the matrix $\underline{\Psi}$ of the scalar products of the input states $\langle \psi_j|\psi_k\rangle = \gamma = e^{-\frac{s}{2}}, \forall j \neq k$.

Demonstration that projectors are solutions of the system of equations can be outlined as follows [16, p. 117].

(i)   from II.2.2a and for $|F_j\rangle \perp |\psi_i\rangle$ $\widehat{T}|F_j\rangle \propto \widehat{R}_j|\psi_j\rangle$

(ii)   from II.2.2b

$$\widehat{R}_i\left(\widehat{T} - \xi_i \widehat{\Psi}_i\right)|F_k\rangle = 0 \xrightarrow{(\langle\psi_i|F_k\rangle = \delta_{ik})} \widehat{R}_i\widehat{R}_k|\psi_k\rangle = \delta_{ik}\widehat{R}_i|\psi_k\rangle$$

(iii) $|\omega_k\rangle = \widehat{R}_k|\psi_k\rangle \rightarrow \widehat{R}_i|\omega_k\rangle = \delta_{ik}|\omega_k\rangle$

(iv)   With $|\Omega\rangle = \sum_k \alpha_k|\omega_k\rangle$, and from (ii) $\longrightarrow \widehat{R}_i\widehat{R}_j|\Omega\rangle = \delta_{ij}|\Omega\rangle$.

As a simple application, we consider an M–ary channel for which we set $\langle\omega_i|\psi_i\rangle = a$, $\forall i$ and $\langle\omega_j|\psi_k\rangle = b$, $\forall i \neq k$. We have $\langle\psi_j|\psi_j\rangle = 1 = \sum_i\langle\psi_j|\omega_i\rangle\langle\omega_i|\psi_j\rangle = a^2 + (M-1)b^2$ and also $\langle\psi_j|\psi_k\rangle = \gamma = \sum_i\langle\psi_j|\omega_i\rangle\langle\omega_i|\psi_k\rangle = 2ab + (M-2)b^2$.

The probability of error can be written as $P_e = \frac{1}{M}\sum_{k=1}^{M}(1-|x_{kk}|^2)$ where $x_{ii} = \langle\omega_i|\psi_i\rangle$ are elements of the matrix $\underline{X}$ for which $\underline{X}^\dagger\,\underline{X} = \underline{\Psi}$. We easily derive

$$P_e = \frac{M-1}{M^2}b^2 = \frac{M-1}{M}\left(\sqrt{1+(M-1)\gamma} - \sqrt{1-\gamma}\right)^2 \quad (II.2.2c)$$

For $M \to \infty$, we again use the method of section II.1.2b where $R = \frac{\log M}{T}$ is the rate of information. We prove that for high intensity $\mu^2 >> R, b^2 \sim \frac{\gamma^2}{4}$, by expansion of the first square root, that the asymptotic form of P_e will be given by $\frac{1}{4}M^{1-2u}, u = \frac{\mu^2}{R}$.

The information performances of this orthogonalized–signal coding channel will be calculated below. For $M = 2$, the probability of error is then given by

$$P_e \equiv p = \frac{1}{2}(1 - \sqrt{1-|\gamma|^2}) \quad (II.2.2d)$$

where $\gamma = \langle \psi_0 | \psi_1 \rangle$ which can easily be derived for $\gamma \in \mathbb{R}$ with the help of Fig. II.1.

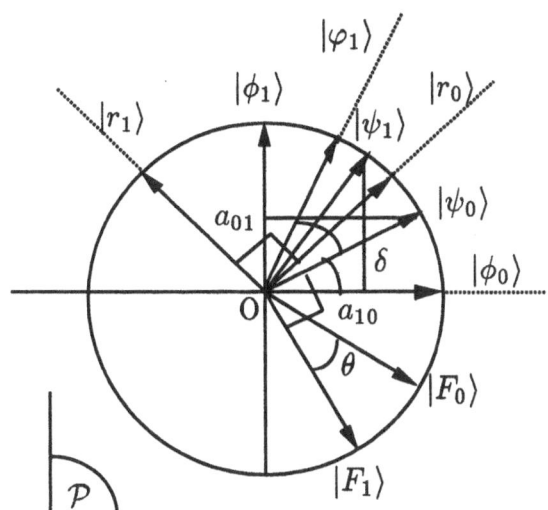

Fig. II. 1. For M=2, a simple geometric representation is helpful. We consider that all states are normed to 1, lying on a plane $\mathcal{P}$. We limit ourselves to given $0 \le \theta \le \frac{\pi}{2}$.

In the plane $\mathcal{P}$ defined by the given input states $(|\psi_0\rangle, |\psi_1\rangle; \theta)$, the states $|\phi_0\rangle$, $|F_0\rangle$ and $|\phi_1\rangle, |F_1\rangle$ can be determined such that $|\phi_0\rangle \perp |\phi_1\rangle$, and $|\psi_0\rangle \perp |F_1\rangle$. In fact, it is easily seen that

$$P_e = \min_{\delta} \frac{1}{2}(a_{10}^2 + a_{01}^2) = \min_{\delta} \frac{1}{2}\left(\cos^2(\theta + \delta) + \sin^2 \delta \right)$$

the minimum is satisfyed for $\widehat{\Pi}_0 = |r_0\rangle\langle r_0|$ and $\delta = \frac{1}{2}(\frac{\pi}{2} - \theta)$, so that $P_e = \frac{1}{2}(1 - \sin \theta)$ with $\cos \theta = \gamma$).

Notice that Kennedy [13], Shapiro [22] proposed an interesting system that can be near–optimum. Basically, that receiver is so that $\widehat{\Pi}_0 = |\psi_0\rangle\langle\psi_0|$ that leads to the channel of Z–type.

For a non orthogonal resolution of identity e.g. $\left(\widehat{\Pi}_0 = |\phi_0\rangle\langle\phi_0|, \right.$ $\widehat{\Pi}_0 = |\varphi_1\rangle\langle\varphi_1|), \langle\varphi_1|\phi_0\rangle = \epsilon,$ such that $\left. 0 \le \epsilon \le \frac{\pi}{2} \right),$ we show that the probability of error becomes $P'_e = \cos^2(\frac{\epsilon+\theta}{2})$ which is larger than P_e.

For $M = 3$, recent calculations [35] suggest a practically feasible system can attain a near–optimum probability of error. That proposed detection scheme is characterized by the set of projections

$$\begin{cases} \hat{\Pi}_1 = |\tilde{w}_1\rangle\langle\tilde{w}_1|, \\ \hat{\Pi}_2(\theta) = |\tilde{w}_2(\theta)\rangle\langle\tilde{w}_2(\theta)|, \\ \hat{\Pi}_3(\theta) = \langle\tilde{w}_3(\theta)|\langle\tilde{w}_3(\theta)| = \mathbb{1} - \hat{\Pi}_1 - \hat{\Pi}_2(\theta), \end{cases} \qquad (\text{II}.2.3a)$$

where $|\tilde{w}_1\rangle = |\alpha_1\rangle$, and

$$|\tilde{w}_2(\theta)\rangle = \frac{\cos\theta}{\sqrt{1-|\gamma_{12}|^2}}(|\alpha_2\rangle - |\alpha_1\rangle\gamma_{12})$$

$$+ \frac{\sin\theta}{\sqrt{1-|c|^2}}\left(\frac{|\alpha_3\rangle - |\alpha_1\rangle\gamma_{13}}{\sqrt{1-|\gamma_{13}|^2}} - c\frac{|\alpha_2\rangle - |\alpha_1\rangle\gamma_{12}}{\sqrt{1-|\gamma_{12}|^2}}\right)$$

$$(\text{II}.2.3b)$$

with

$$c = \frac{\gamma_{23} - \gamma_{12}^*\gamma_{13}}{\sqrt{(1-|\gamma_{12}|^2)(1-|\gamma_{13}|^2)}} \qquad (\text{II}.2.3c)$$

In the above detection scheme, the parameter θ is optimized so that the average error probability is minimized. With a geometrical interpretation, it is easy to see that this scheme specifies a measurement state $|\tilde{w}_1\rangle$ along the state vector $|\alpha_1\rangle$, and chooses the other measurement states $|\tilde{w}_2(\theta)\rangle$ and $|\tilde{w}_3(\theta)\rangle$ on the plane (P) which is perpendicular to the vector $|\alpha_1\rangle$ at its origin. Therefore θ is the angle $(|\alpha_2\rangle_p, |\tilde{w}_2(\theta)\rangle)$, where $|\alpha_2\rangle_p$ is a projection of $|\alpha_2\rangle$ onto the plane (P). Hence, we have : $\Pr(2|1) = \text{Tr}(\hat{\Psi}_1\hat{\Pi}_2(\theta)) = 0$ and $\Pr(3|1) = \text{Tr}(\hat{\Psi}_1\hat{\Pi}_3(\theta)) = 0$.

Furthermore,

$$\Pr(1|1) = \mathrm{Tr}(\widehat{\Psi}_1 \widehat{\Pi}_1) = 1,$$

$$\Pr(2|2) = \mathrm{Tr}(\widehat{\Psi}_2 \widehat{\Pi}_2(\theta)) = \left[1 - |\gamma_{12}|^2\right] \cos^2 \theta,$$

$$\Pr(3|3) = \mathrm{Tr}(\widehat{\Psi}_3 \widehat{\Pi}_3(\theta)) = \left| d^* \sin \theta - \sqrt{1 - |\gamma_{13}|^2 - |d|^2} \cos \theta \right|^2,$$

$$(\mathrm{II.2.3}d)$$

where

$$d = \frac{\gamma_{23} - \gamma_{12}^* \gamma_{13}}{\sqrt{1 - |\gamma_{12}|^2}} \qquad (\mathrm{II.2.3}e)$$

Let us assume that the prior probabilities are equal, i.e. $\xi_i = 1/3 \ \forall i$, then the average error probability is given by

$$\widetilde{P}_e(\theta) = 1 - \frac{1}{3}\left[\Pr(1|1) + \Pr(2|2) + \Pr(3|3)\right]$$

$$= \frac{1}{3}\left[|\gamma_{12}|^2 + |\gamma_{13}|^2 + |d|^2 + 2f \sin^2 \theta + 2g \cos \theta \sin \theta\right]$$

$$= \frac{1}{3}\left[|\gamma_{12}|^2 + |\gamma_{13}|^2 + |d|^2 + f - \sqrt{f^2 + g^2} \cos(2\theta + \phi)\right]$$

$$(\mathrm{II.2.3}f)$$

where

$$f = 1 - |d|^2 - \frac{1}{2}\left(|\gamma_{12}|^2 + |\gamma_{13}|^2\right)$$

$$g = \Re e(d)\sqrt{1 - |\gamma_{13}|^2 - |d|^2}$$

Then the optimum value θ^*, which minimizes (II.2.3f), is easily found such that the minimum average probability of error is given by

$$\widetilde{P}_e = \widetilde{P}_e(\theta) = \frac{1}{3}\left[|\gamma_{12}|^2 + |\gamma_{13}|^2 + |d|^2 - \frac{g^2}{f + \sqrt{f^2 + g^2}}\right] \quad (\mathrm{II.2.3}g)$$

Especially, if $\gamma_{ij} \approx 0 \; \forall i \neq j$, we have $\theta^* \approx 0$ and $d \approx \gamma_{23}$. The average error probability can then be approximated by

$$
\begin{aligned}
\widetilde{P_e} &\approx \frac{1}{3}\left[|\gamma_{12}|^2 + |\gamma_{13}|^2 + |\gamma_{23}|^2\right] \\
&= \frac{1}{3}\left(e^{-|\alpha_1-\alpha_2|^2} + e^{-|\alpha_2-\alpha_3|^2} + e^{-|\alpha_3-\alpha_1|^2}\right)
\end{aligned}
\tag{II.2.3h}
$$

This is only twice as large as the one obtained by the average detection scheme. In this sense, the proposed detection scheme behaves asymptotically like the average error probability processing, and can be calculated very easily.

Notice that because it is often difficult to evaluate the performance of the optimum processor, it is interesting to seek approximate solutions as close as possible to the optimum one. We just discussed "near"–optimality approaches. Now and in analogy to the classical theory, the quantum threshold detector can be considered. Define for $i = 0, 1$

$$
\begin{cases}
E[X|H_i] = \text{Tr}\left(\hat{\rho}_i \hat{\Pi}\right) \\
\hat{\rho}_1(\mu) \sim \hat{\rho}_0 + \frac{\partial}{\partial \mu}\hat{\rho}_1(\mu)\big|_{\mu \to 0}
\end{cases}
\tag{II.2.4a}
$$

where $\hat{\Pi}$ is the Hermitian operator maximizing Δ (II.1.12).

To obtain the equation for $\hat{\Pi}$, we first assume that $\text{Tr}\left(\hat{\rho}_0 \hat{\Pi}\right) = 0$ in the denominator of the expression giving Δ. Then for any Hermitian operator $\left(\hat{\Omega}^\dagger = \hat{\Omega}\right)$ and any real number λ, we require that $\Delta(\hat{\Pi}) = \Delta(\hat{\Pi} + \lambda\hat{\Omega})$ which yield

$$
\text{Tr}\left(2(\hat{\rho}_1 - \hat{\rho}_0)\hat{\Omega} - (\hat{\rho}_0\hat{\Pi} + \hat{\Pi}\hat{\rho}_0)\hat{\Omega}\right) = 0
$$

so that

$$
2\left(\hat{\rho}_1 - \hat{\rho}_0\right) = \hat{\rho}_0\hat{\Pi} + \hat{\Pi}\hat{\rho}_0
\tag{II.2.4b}
$$

It can be seen that the solution is of the form

$$\widehat{\Pi} = 2 \int_0^\infty e^{-\widehat{\rho_0} u} 2 \left(\widehat{\rho_1} - \widehat{\rho_0} \right) e^{-\widehat{\rho_0} u} du \qquad (\text{II.2.4}c)$$

It should be emphasized that this operator is only suboptimal compared with the o.d.o.

The equation (II.2.4b) can also be obtained by using the symmetrical logarithmic derivative of $\widehat{\rho_1}(\mu)$, given by

$$\frac{\partial}{\partial} \widehat{\rho_1}(\mu) \Big|_{\mu \to 0} = \frac{1}{2} \left(\widehat{\rho_0} \widehat{\Pi} + \widehat{\Pi} \widehat{\rho_0} \right) \qquad (\text{II.2.4}d)$$

Helstrom [16, chapter VI] treated in detail the case of detection of coherent state in a thermal state and gave an interesting physical interpretation of the quantum threshold operator. Using the expressions of $\widehat{\rho_0}$ and $\widehat{\rho_1}$ (see II.3.5), it is established

$$\begin{cases} \frac{\partial}{\partial \mu} \widehat{\rho_1}(\mu) \Big|_{\mu \to 0} = \frac{1}{\langle N \rangle} \left(\widehat{a} \widehat{\rho_0} + \widehat{\rho_0} \widehat{a}^\dagger \right) \\ \widehat{a} \widehat{\rho_0} = e^{-w} \widehat{\rho_0} \widehat{a} \\ \widehat{\rho_0} \widehat{a}^\dagger = e^{-w} \widehat{a}^\dagger \widehat{\rho_0} \\ \\ \widehat{\Pi} = \frac{1}{\langle N \rangle + \frac{1}{2}} \left(\widehat{a} + \widehat{a}^\dagger \right) \end{cases} \qquad (\text{II.2.4}e)$$

It is important to notice that this quantum threshold operator is proportional to the position operator $\widehat{a}_1$ defined in (I.3.1) for the coherent detection. However, this is valid only for $\mu \to 0$. For higher values of μ, the formula (II.2.4c) is more appropriate but the interpretation in terms of the criterion (II.1.12) is no longer valid. Detailed analysis and application of this operator to various cases have been presented by Yoshitani [9].

Another application that is the quantum phase state, is given below.

II.2.2 Application to the Quantum Phase State

As has been done for the detection criterion, some elements of the quantum estimation theory are now recalled bearing in mind an application to estimate a quantum phase [16]. Suppose we want to *estimate* a phase $\phi \in]-\pi, \pi[$. To treat this ptoblem, let the input field defined by $\{\hat{a}_0, |\psi\rangle_0\}$ and $\hat{\Phi}$ be some shift operator. Let the shifted field given by $\{\hat{a}, |\psi\rangle\}$ and $d\hat{\Pi}(\phi)$ be the p.o.m. so that

$$
\begin{cases}
|\psi\rangle \equiv \exp(i\hat{\Phi}\hat{a}_0^\dagger \hat{a}_0)|\psi\rangle_0 \\[2mm]
p(\phi|\hat{\Phi})d\phi = \langle\psi|d\hat{\Pi}(\phi)|\psi\rangle \\[2mm]
d\hat{\Pi}(\phi) = d\hat{\Pi}(\phi)^\dagger \\[2mm]
\mathbb{1} = \int_{\phi=-\pi}^{\pi} d\hat{\Pi}(\phi)
\end{cases}
\tag{II.2.5a}
$$

and we seek $d\hat{\Pi}(\phi)$. Setting $|\psi\rangle_0 = |\psi_i| \exp(i\chi_0)$ and $|\exp(i\phi)\rangle \equiv \sum_{n=0}^{\infty} \exp(in\phi)|n\rangle$, we obtain

$$
d\hat{\Pi}(\phi) = |\exp(i\phi), |\psi\rangle\rangle\langle\exp(i\phi), |\psi\rangle|
\tag{II.2.5b}
$$

To prove this result, some basic equations are briefly recalled. Let $\theta \in \Theta$ be the parameter to be estimated. Its probability density *a priori* is given by $z(\theta)$, and let $\tilde{\theta}$ be the estimate. The p.o.m is denoted by $d\hat{\Pi}(\tilde{\theta})$. The cost is $C(\tilde{\theta}, \theta) = -\delta(\tilde{\theta} - \theta)$ whose mean is

$$
\mathrm{Tr} \int_\Theta \hat{W}(\tilde{\theta}) d\hat{\Pi}(\tilde{\theta})
$$

where $\hat{W}(\tilde{\theta}) = \int_\Theta \rho(\theta) z(\theta) C(\tilde{\theta}, \theta)$ is the risk operator.

Using the Lagrange multipliers method, we set the equations of the optimization with $z(\theta) = \frac{1}{2\pi}$

$$\begin{cases} \left[\widehat{\varGamma} - \widehat{W}(\tilde{\theta})\right]d\widehat{\varPi}(\tilde{\theta}) = 0 \\ \widehat{\varGamma} - \widehat{W}(\tilde{\theta}) \geq 0 \\ \widehat{\varGamma} = \int_{\Theta} \widehat{W}(\tilde{\theta})d\widehat{\varPi}(\tilde{\theta}) \end{cases} \qquad (\text{II.2.5}c)$$

defined so that $\widehat{\varGamma}^{\dagger} = \widehat{\varGamma}$. The p.o.m. $d\widehat{\varPi}(\phi)$ thus derived will be called *optimal* under the maximum likelihood ratio criterion. Here we obtain

$$d\widehat{\varPi}(\tilde{\theta}) = \xi(\tilde{\theta})d\tilde{\theta}; \ \xi(\tilde{\theta}) \geq 0 \ ; \ \int_{-\pi}^{\pi} \xi(\tilde{\theta})d\tilde{\theta} = 1$$

Now under a unitary transformation, the density operator of the system will be given by $\rho(\theta) = \exp(-i\widehat{N}\theta)\rho_0 \exp(i\widehat{N}\theta)$, we have from (II.2.5c)

$$\begin{cases} [\widehat{\varGamma} - \rho(\tilde{\theta})]\xi(\tilde{\theta}) = 0 \\ \widehat{\varGamma} - \rho(\tilde{\theta}) \geq 0 \\ \widehat{\varGamma} = \int_{-\pi}^{\pi} \xi(\tilde{\theta})\rho(\tilde{\theta})d\tilde{\theta} \end{cases} \qquad (\text{II.2.5}d)$$

defined so that

$$\begin{cases} \xi(\tilde{\theta}) = \exp(-i\widehat{N}\tilde{\theta})\xi_0 \exp(i\widehat{N}\tilde{\theta}) \\ \xi_0 = \frac{1}{2\pi}|\gamma\rangle\langle\gamma| \end{cases} \qquad (\text{II.2.5}e)$$

where $\widehat{N}|k\rangle = k|k\rangle$. The state $|\gamma\rangle$ is defined by its coordinates $\{\gamma_k\}$ in the space spanned by $\{|k\rangle\}$, with

$$\gamma_k = \langle k|\gamma\rangle = \frac{x_k}{|x_k|} = \begin{cases} \exp(i\chi_k), & x_k \neq 0 \\ 1, & x_k = 0 \end{cases}$$

Finally the posterior probability density function of the estimate $\tilde{\theta}$ of the displacement θ is given by

$$q(\tilde{\theta}|\theta) = \mathrm{Tr}\ \left(\rho(\theta)\xi(\tilde{\theta})\right) \equiv q(\phi) \qquad (\mathrm{II}.2.5f)$$

where $q(\phi) = \langle\psi|\xi(\phi)|\psi\rangle$ with $\phi = \tilde{\theta} - \theta$ and $\rho_0 = |\psi\rangle\langle\psi|$.

Extension of the results and methods to continuous input and output values seems difficult to perform. For detailed study of quantum detection and estimation problems, one can refer to the books [16,17,25].

As the capacity criterion has been extensively studied in several textbooks (see [3,5] for instance) and briefly recalled in the Sect. II.1.2b for the classical formulation, we focus in the next section on the cut–off rate criterion. Classical expressions that are applicable to photon counting communication channel have been published [21,23,24]. Finally, a brief application of the rate distortion function is given.

II.2.3 Communication with M Coherent States

We present exact results for the noiseless channel and some approximate calculations for channels with thermal noise of weak power.

II.2.3a Noiseless Channels

By analogy to the classical channel, it is possible to define the mutual information in the form

$$\mathrm{I}(\Xi,\hat{\rho}_j;\hat{R}_k) = S\left(\sum_j \xi_j q(k|j)\right) - \sum_j \xi_j S\left(q(k|j)\right) \qquad (\mathrm{II}.3.1a)$$

where $\Xi = \{\xi_j\}_0^{M-1}$ is the vector of input *a priori* probability and

$$q(k|j) = \text{Tr} \left(\widehat{\rho}_i \widehat{R}_k \right) \qquad (III.3.1b)$$

is the transition probability: $\forall j$, $q(k|j) \geq 0$, $\sum_k q(k|j) = \xi_j$, $\sum_k q(k|j) = 1$.

Because of the nonlinear form of the entropy function S, it is, in general, very difficult to determine the output measurement operator $\widehat{R}_k$ that maximizes (II.3.1a) for given input states $\widehat{\rho}_j$. However, an important result demonstrated by Holevo [12,14] states that

For given ξ_j and $\{\widehat{\rho}_j\}$, the mutual information given by (II.2.4a) is upperbounded so that

$$I(\Xi, \widehat{\rho}_j; \widehat{R}_k) \leq S \left(\sum_{j=0}^{M-1} \xi_i \widehat{\rho}_j \right) - \sum_{j=0}^{M-1} \xi_j S(\widehat{\rho}_j) \qquad (II.3.2)$$

the equality being reached for $\widehat{\rho}_j$ pairwise commuting.

The demonstration is based on Jensen's inequality generalized to operators. Extension of this theorem to a countable number of input symbols $M \to \infty$, and a separable Hilbert space ($d \to \infty$) has recently been proposed by Hall and O'Rourke [34]. This approach closely follows the classical theory but using an undefined measurement operator, can be considered as semi–classical.

The same approach is developed in the next section for the study of the cut off rate.

II.2.3b Cut–Off Rate Criterion

An information–theoretic criterion, that is, the cut–off rate, has interesting properties, as already pointed out (see § II.1.2c), and is used in the following in the form

$$R_0 = -\min_{\mathcal{S}} \log\left(J_0\left(\Xi, \{\rho_i\}; \{\hat{R}_j\}\right)\right)$$

(II.3.3a)

$$= -\min_{\mathcal{S}} \log\left(\sum_{k,l} \xi_k \xi_l \sum_j \sqrt{p_{kj}p_{lj}}\right)$$

where $\mathcal{S}$ is the convex set

$$\mathcal{S} = \begin{cases} \forall j, \ \hat{\rho}_j \geq 0, \operatorname{Tr}(\hat{\rho}_j) = 1 \\ \xi_j \geq 0, \sum_j \xi_j = 1 \\ \forall k, \ \hat{R}_k \geq 0, \sum_k \hat{R}_k = \mathbb{1} \end{cases}$$

(II.3.4)

We follow the method [20] and also refer to relevant recent work [31] to prove that the cut–off rate is maximized by a p.o.m. of rank one. The proof can be sketched, based on the following properties of the function operator J_0

(i) For given ξ_j and $\{\rho_j\}$, $J_0\left(\{\rho_j\}, \{\hat{R}_k\}\right)$ is a convex $\cup$ function with respect to the $\hat{R}_k$

(ii) Substitution of two arbitrary p.o.m., for example, $\hat{R}'_{N-2} = \hat{R}_{N-1} + \hat{R}_{N-2}$ with $\hat{R}'_k = \hat{R}_k$ for $j \in [1, N-3]$ leads to $R'_0 \leq R_0$.

(iii) For a $\mathcal{H}_d$ Hilbert space of dimension d, the solution is $\hat{R}_k = r_k|r_k\rangle\langle r_k|$, such that $r_k \leq 1, \sum_k r_k = d, \|r_k\| = 1$.

(iv) $d^2 \geq N \geq d$.

The result (iv) means that the information rate can be larger than $\log d$ which is the maximum amount of information derived from a classical consideration.

Let us apply these results to an important example: the input states are M–amplitude modulated coherent states

$$\{\rho_l\}_{l=0}^{M-1} = \left|\mu\frac{2l - M + 1}{M - 1}\right\rangle\left\langle\mu\frac{2l - M + 1}{M - 1}\right| = |\alpha_l\rangle\langle\alpha_l| \qquad (\text{II}.3.5)$$

with $\mu \in \mathbb{R}^+$ and (II.3.3,4) are calculated with $\widehat{R}_k$ as projectors.

The principle of the calculation is based on the well–known method of minimization of a quadratic form. In fact, introducing the matrix $\underline{T}$ whose elements are $t_{kl} = \sqrt{p_{kl}}$, it is clear that II.3.3a takes the form $-\min_{S}\left(\Xi^+\underline{T}\,\underline{T}^+\Xi\right)$. The solution of $L = \min_{S}\Xi^+\underline{G}\Xi$ is obtained by means of usual Lagrange multipliers. The $\underline{G} = \{\langle\alpha_l|\alpha_m\rangle\}$ is the Gram matrix of the M^2 scalar products of Eq. (II.3.5), and S is the convex set (II.3.4).

For discrete input states and continuous ouput (e.g. measurement of the operator $\widehat{a}_1$ (see II.1.30)), the expression of R_0 takes the form

$$R_0 = -\min_{S}\log\left(\sum_{k,l}\xi_k\xi_l\int_{x\in X}\sqrt{p_k(x)p_l(x)}dx\right) \qquad (\text{II}.3.3b)$$

Therefore

$$R_0 = -\log\left(\frac{\det \underline{G}}{\sum_{kl}\omega_{kl}}\right) \text{ such that } \forall k, \ q_k = \sum_{l}\omega_{kl} > 0$$

where $\det \underline{G}$ denotes the determinant (> 0) of the definite and non-negative matrix, ω_{kl} are elements of $\underline{G}^{-1}$. For one or more than one $(m, n, ...)$ such that $q_{m,n,...} \leq 0$, it is arbitrarily imposed $\xi_{m,n,...} = 0$. A detailled analysis is given in [28].

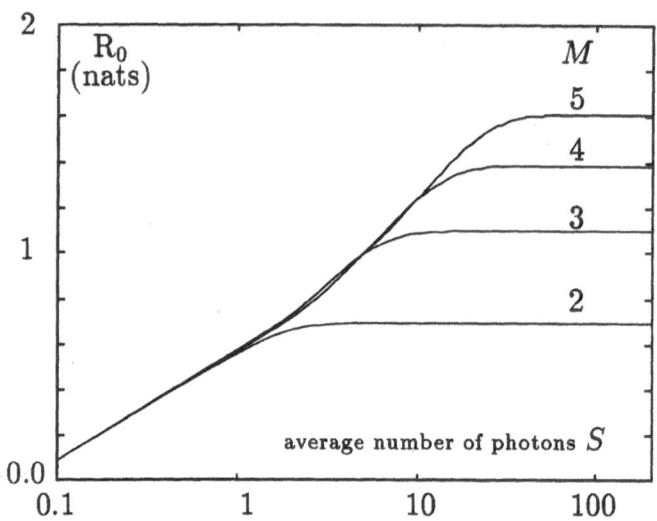

Fig. II.2. Cut–off rate versus the average number of coherent signal photons $S = |\mu|^2$, for a noiseless channel : M amplitude–modulated coherent states with M=2,...5.

From the previous figure, it can be noticed that for $S = \mu^2 \leq 1$, it is no interest *encoding* the source with $M > 2$.

Now, we would like to treat the case $M = N = 2$ using the method pointed out in Sect. II.2.1. We set $p(1|1) = p = \cos^2 \delta, p(2|2) = q = 1 - \cos^2(\theta + \delta), \gamma = \langle \alpha_0 | \alpha_1 \rangle = \cos \theta$ and we calculate the cut–off rate

$$R_0^{(2)}(\delta, \theta) = \log \frac{2}{1 + \sqrt{p(1 - q)} + \sqrt{q(1 - p)}} \qquad (\text{II.3.6})$$

for given $0 \leq \theta \leq \frac{\pi}{2}$. We seek δ^* that optimizes R_0. There is a symmetrical axis $\delta = \frac{\pi}{4}$ so that $\delta^* = \frac{\pi}{4} - \frac{\theta}{2}$ which is the value that minimizes the probability of error in detection and characterizes the o.d.o., as is already seen from the geometrical interpretation (cf. Fig. II.1).

For $M = 2, N = 3$, a strategy that uses a third projector located in the plane defined by the two input pure states leads to a degradation of performance $R_0^{(3)}(\delta, \theta) < \log(2), \forall \delta, \theta$ as can be seen by using the generalized measurements $\frac{2}{3}|r_i\rangle\langle r_i|$, such that $\sum_i |r_i\rangle\langle r_i| = \mathbb{1}$, $\langle r_i|r_j\rangle \neq \delta_{ij}, i \neq j = 1, 2, 3$. Notice that the $\min_\delta P_e$ is obtained for $\delta^* = \frac{\pi}{3} - \frac{\theta}{2}$, where $\frac{\pi}{6} \leq \theta \leq \frac{2\pi}{3}$. We set $p(1|1) = p(2|1) = p = \frac{1}{2}\cos^2 \delta, p(1|2) = p(2|2) = q = \frac{1}{2}\cos^2(\theta + \delta), p(3|1) = r = \sin^2 \delta, p(3|2) = w = \sin^2(\theta + \delta)$, we then calculate

$$R_0^{(3)}(\delta, \theta) = \log \frac{2}{1 + \sqrt{rw} + \sqrt{p(1 - q - w)} + \sqrt{q(1 - p - r)}}$$

$$(\text{II.3.7})$$

This result can be generalized for arbitrary M, N in a particular case [15], for which we have

$$\begin{cases} p(j|j) = 1 - (M-1)p - (N-M)r, & \\ p(k|j) = p & j, k = 1, ..., M & k \neq j, \\ \qquad = r & k = M+1, ..., N & k \neq j \end{cases}$$

It is simple to calculate

$$R_0^{(N)} = \log \frac{M}{D(N, M, r, p)}$$

$$D(N, M, r, p) = 1 + (M-1)(N-M)r$$

$$+ 2(M-1)\sqrt{p\left(1 - (M-1)p - (N-M)r\right)}$$

$$(\text{II.3.8})$$

which can be applied to M-ary channel with an erasure symbol. Analysis of Eq. (II.3.8) versus N and M for given p show that for a special case of $N = M$ and $p = r$, communication with two real amplitude coherent states of weak intensity $\mu^2 \leq \frac{\log M}{2}$, degrades with increasing M [28].

However, for $M = 2, N = 3$, performance of a binary symmetric erasure channel with classical detection and optimized threshold can be made higher than those of $M = N = 2$ [5, p. 400] and we also may calculate the mutual information and then the capacity

$$C = (1 - \epsilon) \log \frac{M}{1 - \epsilon} + (M - 1)b^2 \log b^2 + a^2 \log a^2 \qquad \text{(II.3.9)}$$

where a^2 and $b^2 = \frac{1}{M^2} \left(\sqrt{1 + (M - 1)\gamma} - \sqrt{1 - \gamma} \right)^2$ are the exact and error probability respectively $(a^2 + (M - 1)b^2 + \epsilon = 1)$, and $\gamma = e^{-\mu^2 T}$, μ is the real amplitude of the light detected within a time interval T.

For $M \to \infty$, we use once more the method of Sect. II.1.2b, where $R = \frac{\log M}{T}$ is the rate of information and $u = \frac{\mu^2}{R}$ to prove that the capacity is such that

$$0 < u < 1, \quad b^2 \sim \frac{\gamma}{M}, \quad C \sim (1 - \gamma) \log M \to \infty$$

$$u \geq 1, \qquad b^2 \sim \frac{\gamma^2}{4}, \qquad C \sim \log M \to \infty$$

and the cut–off rate will be given by

$$R_0 = \log \frac{M}{M - 1 + \epsilon - (M - 2)a^2 + 2(M - 1)\sqrt{a^2 b^2}} \qquad \text{(II.3.10)}$$

plotted in the next figure for several values of M.

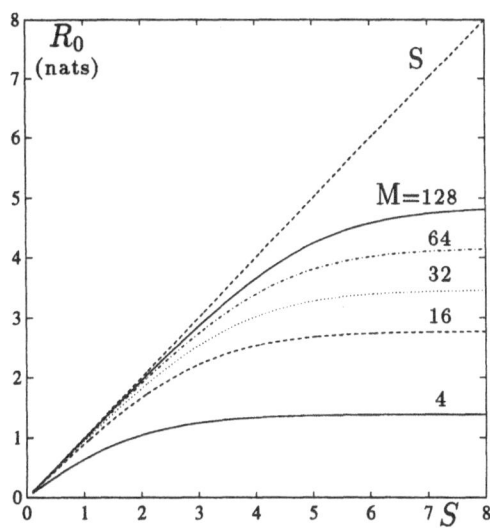

Fig. II.3. Cut–off rate performances for M-ary orthogonal signal coding quantum channel for $M = 4$, 16, 32, 64, 128 versus S for $\epsilon = 0$, $(a^2 = 1-(M-1)b^2)$, versus the average input signal photon number $S = \mu^2 T$. The asymptotic curve is quoted S and is given by $R_0 = -\log\gamma = S$. It is observed that we have $R_0(M,S) \leq S$, $\forall M, S$.

II.2.3c Rate–Distortion

As in Sect. II.1.2d, the joint input–output probability, the input probability and the output probability are rewritten here in terms of the $\{\hat\rho_j\}$ input density operators and the p.o.m. $\{\hat R_k\}$

$$
\begin{cases}
\pi_{jk} = \text{Tr}\xi_j\hat\rho_j\hat R_k \\
p_j = \sum_k \pi_{jk} = \xi_j \\
r_k = \sum_j \pi_{jk} = \text{Tr}\,\hat\rho\hat R_k \\
\hat\rho = \sum_j \xi_j\hat\rho_j = \sum_j \xi_j|\psi_j\rangle\langle\psi_j|, \left(\xi_j \geq 0, \sum_j \xi_j = 1, \text{Tr}\hat\rho_j = 1, \forall j\right) \\
\sum_k q(k|j) = \sum_k \text{Tr}\hat\rho_j\hat R_k = 1, \forall j, \left(\sum_k \hat R_k = 1\!\!1\right)
\end{cases}
$$

$$(\text{II.3.11})$$

We then need to define the measure of distortion. It is taken here again as $\Delta_{jk} = 1 - \delta_{jk}$ (see Sect. II.1.2b), where we assume that $j, k = 1, ..., M$. Its average value is given by

$$\langle \widehat{\Delta} \rangle = \sum_{j,k} \pi_{jk} \Delta_{jk} = 1 - \mathrm{Tr} \sum_j \xi_j \widehat{\rho}_j \widehat{R}_j = P_e \overset{\mathrm{def}}{=} D \qquad (\mathrm{II.3.12})$$

As previously, the mutual information is

$$\begin{aligned}
\mathrm{I}(\widehat{\rho}_j; \widehat{R}_k) &= \sum_{j,k} \pi_{jk} \log \frac{\pi_{jk}}{p_j r_k} \\
&= \sum_{j,k} \pi_{jk} (\log \pi_{jk} - \log r_k) - \sum_j \xi_j \log \xi_j
\end{aligned} \qquad (\mathrm{II.3.13})$$

To give (II.3.13) a different form, we notice

$$\mathrm{S}(\widehat{\Pi}) = -\mathrm{Tr} \left(\widehat{\Pi} \log \widehat{\Pi} \right) = -\sum_{jk} \pi_{jk} \log \pi_{jk}$$

where $\pi_{jk} = \langle \psi_j, \chi_k | \widehat{\Pi} | \psi_j, \chi_k \rangle$, $\widehat{\Pi} = \{ \xi_j \widehat{\rho}_j \widehat{R}_k \}$, $(\sum_{jk} \pi_{jk} = 1)$. Therefore Eq. (II.3.13) can take the form

$$\mathrm{I}(j|k) = \mathrm{H}(\Xi) + \mathrm{Tr} \left(\widehat{\Pi} \left(\log \widehat{\Pi} - \log \widehat{R}_k \right) \right) \qquad (\mathrm{II.3.14})$$

where $H(\Xi) = -\sum_j \xi_j \log \xi_j$.

Then the rate distortion can be rewritten as solution $(\widehat{R}_k)$ to the problem of optimization for given Ξ

$$\begin{cases}
R(D) = \min_{\widehat{R}_k} \left[H(\Xi) + \mathrm{Tr} \left(\widehat{\Pi} \left(\log \widehat{\Pi} - \log \widehat{R}_k \right) \right) \right] \\
D = \mathrm{Tr} \left(\widehat{\Pi} \widehat{\Delta} \right)
\end{cases} \qquad (\mathrm{II.3.15a})$$

that can be solved by using the standard Lagrange multipliers method. Here, because $D = P_e$, it is simpler to make use of Eq. (II.1.26b) and show that the expression of the p.o.m. $\Pi = \{\widehat{R}_k\}$ is such that

$$\mathrm{Tr}\left(\widehat{R}_k(\xi_j \widehat{\rho}_j - \widehat{K}_k)\right) = 0 \qquad (\text{II.3.15}b)$$

given $\widehat{K}_k = \sum_j \xi_j \widehat{\rho}_j e^{s\Delta_{jk}} \mu_j$ and $\mu_j = \sum_k w_{jk}, w_{jk} = \{A^{-1}\}_{jk}$. From Eq. (II.3.15b), it follows that

$$\widehat{R}_j = \frac{d_M}{D - d_M}\, \widehat{\rho}^{-1}\left(\xi_j \widehat{\rho}_j - \frac{MD}{d_M}\, \widehat{\rho}\right) \qquad (\text{II.3.15}c)$$

where $d_M = \frac{M-1}{M}$ and D is simply related to the scalar product of the input states (see II.2.2c). As we consider here pure states, the inverse of $\widehat{\rho}$ exists for non commuting pairs $[\widehat{\rho}_j, \widehat{\rho}_k] \neq 0$ (Sect. I.1.1a).

We conclude this section with some explicit expressions of the rate distorsion functions computed for different receivers in a binary channel: $\Xi = (\xi, 1 - \xi)$. Actually, except for the o.d.o., the mutual information is not minimised subject to a given distorsion measure.

(i) For example, given $\widehat{R}_k = \widehat{N}$ and for $0 \leq D < 1 - \xi$, we obtain [19]:

$$I_N(D) = D\log(D) - (1-\xi)\log(1-\xi) - (D+\xi)\log(D+\xi) \qquad (\text{II.3.16}a)$$

By noting that for $\xi, D \leq \frac{1}{2}$, we have

$$\begin{cases} -(1-\xi)\log(1-\xi) - (D+\xi)\log(D+\xi) \leq -(1+D)\log\frac{1+D}{2} \\ D\log D \leq D(D-1) \\ -(1+D)\log(1+D) \leq -D(D+1) \end{cases}$$

$$(\text{II.3.16}b)$$

A very simple upper–bound can be written as

$$I_N(D) \le (1+D)\log 2 - 2D \qquad\text{(II.3.16c)}$$

which appears reasonably good for $\xi \to \frac{1}{2}$. A lower bound seems more difficult to obtain. We found $I_N(D) \ge -\frac{1}{e} + \xi + (\frac{D}{1-\xi})\xi \log \xi$. This bound is rather poor but appears useful for some values of $\xi \sim 0.2$.

(ii) As a second example, we consider the homodyne operator $\widehat{R}_k = \widehat{X}$ for which the measured signal is a continuous variable:

$$I_X(D) = H(p) - H(r) \quad\text{for } 0 \le D < \frac{1}{2} \qquad\text{(II.3.17)}$$

where $H(p) = p\log(p) + (1-p)\log(1-p)$ and $r = p + \xi(1-2p)$. The probability $p = \text{Prob}(k = 0|j = 1) = \text{Prob}(k = 1|j = 0)$, is given so that $p = D = \frac{1}{2}(1 - Q(\sqrt{\frac{S}{2}}))$ where $Q(u)$ is the error function.

(iii) The third example is the o.d.o. completely defined by

$$\begin{cases}
\lambda = \frac{\xi}{1-\xi}; \ h = 1 - e^{-S}; \ \Lambda = \sqrt{\lambda h + \frac{(1-\lambda)^2}{4}}; \ \eta = \Lambda + \frac{1-\lambda}{2} \\
p = \frac{\eta-h}{2\Lambda}; \ q = \frac{\eta+\lambda h}{2\Lambda}; \ r_0 = \xi(1-p) + (1-\xi)(1-q) \\[4pt]
D = \xi p + (1-\xi)(1-q) \\
I_o(D) = \xi \left((1-p)\log\frac{1-p}{r_0} + p\log\frac{p}{1-r_0} \right) \\
\qquad\quad + (1-\xi)\left((1-q)\log\frac{1-q}{r_0} + q\log\frac{q}{1-r_0} \right)
\end{cases}$$

$$\text{(II.3.18)}$$

where we set $S = |\langle \psi_1|\psi_0\rangle|^2$ and $|\psi_0\rangle, |\psi_1\rangle$ are the pure input states. Because the distorsion criterion is of a probability of error type, it is clear that $I_o(D) = R(D)$, $R(D)$ being the classical rate distorsion function (II.1.26).

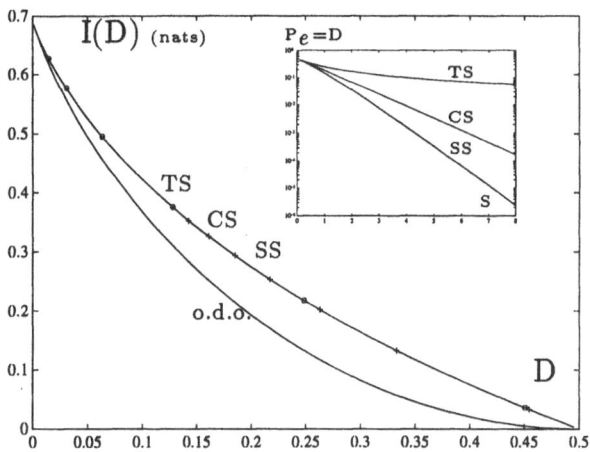

Fig. II.4. Mutual information $I(D)$ for binary channel in the case of a priori probability $x = 0.5$ for fixed probability of error $P_e = D$ for photoncounting channel (N). Notice that the three types of light, coherent (CS), squeezed (SS) and thermal (TS), yield same $I(D)$ although the P_e are different. The (o.d.o.) curve calculated for optimum detection operator with eqs. II.3.18, corresponds to the rate distorsion function given by eqs. II.1.26.

Now, due to the fact that the rate distorsion is a strictly monotonic decreasing function and as $I_N(0) = R(0) = H(\xi)$ and because $D_M^N > D_M^{odo}$ (D_M^r the maximum value of the average value of the distorsion for a receiver r), we can prove that $I_N(D) \geq R(D)$. Therefore, it is observed that higher complexity encoder than the photocounting scheme is required to achieve a performance closer to the $R(D)$. This is due to the fact the photocounting channel is of Z–type that is to say a non–symmetrical one.

A more detailed version of these results will be published in a future work [36].

II.2.3d Channel with Thermal Noise : Approximate Results

Let us consider M pure coherent states described by the density operators $\hat{\rho}_i = |\mu_i\rangle\langle\mu_i|$ where μ_i is the complex envelope of the input field given with equal *a priori* probability ξ_i, $(i = 0, ..., M-1)$. We are also given the scalar products $\gamma_{km} = \langle\mu_k|\mu_m\rangle$. For noiseless channels, the orthogonal projectors of rank one $\widehat{\Pi}_j = |\omega_j\rangle\langle\omega_j|$ with the resolution of identity $\sum_{j=0}^{M-1} \widehat{\Pi}_j = \mathbb{1}_M$ are the optimal measurement operators to minimize the probability of error [16, p.98]. For binary channels with thermal noise of weak average photon number N_0, in order to solve the detection equation $(\hat{\rho}_1 - \lambda\hat{\rho}_0)|\eta\rangle = \eta|\eta\rangle$, some approximate results have been established in terms of the function $F(\beta^*) = e^{\frac{|\beta|^2}{2}}\langle\beta|\eta\rangle$. The linear equation satisfied by $F(\beta^*)$ is expanded in $v = \frac{N_0}{1+N_0}$ and it is shown that up to the first order, the false–alarm and detection probabilities can be approximated [9].

Another method, based on the modification of the resolution of identity [16, p. 184] and of simple interpretation, is developed below. In the presence of thermal noise, the density operators [33] are first expanded under the $P(\alpha)$–representation [1,2]

$$\hat{\rho}_i = \frac{1}{\pi N_0} \int_{-\infty}^{\infty} \exp\left(-\frac{|\alpha - \mu_i|^2}{N_0}\right) |\alpha\rangle\langle\alpha| d^2\alpha \qquad (\text{II.3.19}a)$$

where $|\alpha\rangle\langle\alpha| = D(\alpha)|0\rangle\langle0|D^\dagger(\alpha)$. These states are then projected on the coherent states $\{|\mu_i\rangle\}$ in order to evaluate

$$\langle \mu_k | \widehat{\rho}_i | \mu_\ell \rangle = \frac{1}{1 + N_0}$$

$$\exp\left(\frac{N_0}{1 + N_0} \mu_k^* \mu_\ell + \frac{\mu_k^* \mu_i + \mu_i^* \mu_m - |\mu_i|^2}{1 + N_0} - \frac{|\mu_k|^2}{2} - \frac{|\mu_\ell|^2}{2} \right)$$

$$(\text{II}.3.19b)$$

Therefore, the measurement process involves a resolution of identity on the infinite dimensional Hilbert space as required by (II.3.4) that is a difficult problem to solve. However, some useful bounds are easy to obtain.

In fact, as was done for the detection problem, we assume that the new measurement operators are approximately given by

$$\begin{cases} \widehat{\Omega}_j = \widehat{\Pi}_j + \frac{1}{M} \widehat{\Pi}'_j \\ \sum_{j=0}^{M-1} \widehat{\Omega}_j = \mathbb{1}_M \end{cases} \qquad (\text{II}.3.20a, b)$$

Now, it is useful to expand the detection operator on the input states to read

$$|\omega_\ell\rangle = \sum_{j=0}^{M-1} a_{\ell j} |\mu_j\rangle \qquad (\text{II}.3.21a)$$

so that for the probability distribution

$$p_{ij} = \text{Tr}\left(\widehat{\rho}_i \widehat{\Omega}_j \right) = \langle \omega_j | \widehat{\rho}_i | \omega_j \rangle + \frac{1}{M} \left(1 - \sum_{\ell=0}^{M-1} \langle \omega_\ell | \widehat{\rho}_i | \omega_\ell \rangle \right)$$

$$(\text{II}.3.21b)$$

where

$$\langle \omega_m | \widehat{\rho}_i | \omega_m \rangle = \sum_{k\ell} a_{m\ell} a_{mk}^* \langle \mu_k | \widehat{\rho}_i | \mu_\ell \rangle \qquad (\text{II}.3.21c)$$

As shown [26], the function J_0 defined by $R_0 = -\min_{\xi_i} \log(J_0)$ is upper bounded in the following manner

$$- \quad \mathrm{J}_0^u \geq \sum_{k,\ell=0}^{M-1} \xi_k \xi_\ell \left| \int \sqrt{P_k(\mu)P_\ell(\mu)} d^2\mu \right| \qquad (\text{II.3.22}a)$$

$$- \quad \mathrm{J}_0^v \geq \sum_{\substack{k,\ell=0, \\ k \neq \ell}}^{M-1} \xi_k \xi_\ell \sqrt{\mathrm{Tr}\left[\hat{\rho}_k \hat{\rho}_\ell\right]} + \sum_{k=0}^{M-1} \xi_k^2 \qquad (\text{II.3.22}b)$$

where $P_k(\mu) = \frac{1}{\pi N_0} \exp\left(-\frac{|\alpha-\mu_k|^2}{N_0}\right)$.

Demonstration of these bounds is based on the $P(\alpha)$ representation and the spectral decomposition of the density operator.

$$J = \sum_{k,\ell=0}^{M-1} \xi_k \xi_\ell J_{k\ell}, \quad J_{k\ell} = \sum_{n=0}^{N-1} \sqrt{\mathrm{Tr}\left(\hat{\rho}_k \widehat{R}_n\right) \mathrm{Tr}\left(\hat{\rho}_\ell \widehat{R}_n\right)},$$

(i) $\quad \hat{\rho}_k = \int P_k(\mu)|\mu\rangle\langle\mu|d^2\mu$

(ii) $\quad J_{k\ell} \geq \left| \int \sqrt{P_k(\mu)P_\ell(\mu)} \sum_n \langle\mu|\widehat{R}_n|\mu\rangle d^2\mu \right|$

$\qquad = \left| \int \sqrt{P_k(\mu)P_\ell(\mu)} d^2\mu \right| \rightarrow (\text{II.3.22}a)$

(i) $\quad k = \ell, \quad \sum_n \widehat{R}_n = 1, \ J_{kk} = 1 \rightarrow J = \sum_k \xi_k^2.$

(ii) $\quad k \neq \ell, \quad \hat{\rho}_k = \sum_j \lambda_j |\omega_j^{(k)}\rangle\langle\omega_j^{(k)}|$

$\qquad \mathrm{Tr}\left(\hat{\rho}_k \widehat{R}_n\right) = \sum_j \lambda_j \Omega_j^{(kn)} \Omega_j^{(\ell n)*} \geq 0,$

$\qquad \left(\Omega_j^{(kn)} = \langle\omega_j^{(k)}|\widehat{R}_n|\omega_j^{(k)}\rangle\right),$

$\qquad J_{k\ell} \geq \sqrt{\sum_n \mathrm{Tr}\left(\hat{\rho}_k \widehat{R}_n\right) \mathrm{Tr}\left(\hat{\rho}_\ell \widehat{R}_n\right)} = \sqrt{\mathrm{Tr}\left(\hat{\rho}_k \hat{\rho}_\ell\right)}$

(iii) $\quad$ (i)+(ii) $\rightarrow$ (II.3.22b)

Both bounds are useful, the domain of their interest depend only on the function $f(S, N_0) = \max \left(e^{-\frac{S}{N_0}}, \frac{1}{\sqrt{1 + 2N_0}} e^{-\frac{S}{\sqrt{1 + 2N_0}}} \right)$, where $\mu^2 = S$.

Let us illustrate the calculations by an application to the noisy binary channel. It is emphasized again that the equation of optimization is not solved. Only approximate results are calculated. We find the transition probabilities for this binary symmetrical channel (OOK modulation $\mu_0 = 0, \mu_1 = \mu, \gamma^2 = e^{-S}$), e.g.

$$
\begin{aligned}
p_{10} &= \frac{1}{2} \left[1 + \frac{1}{\sqrt{1 - \gamma^2}} \left(\langle \mu | \hat{\rho}_1 | \mu \rangle - \langle 0 | \hat{\rho}_1 | 0 \rangle \right) \right] \\
&= \frac{1}{2} \left[1 + \frac{1}{(1 + N_0)\sqrt{1 - e^{-S}}} \left(e^{-\frac{S}{1 + N_0}} - 1 \right) \right] \qquad \text{(II.3.23)} \\
&= \frac{1}{2}(1 + \beta)
\end{aligned}
$$

where the expressions $\langle 0 | \hat{\rho}_1 | 0 \rangle = \frac{1}{1 + N_0} e^{-\frac{S}{1 + N_0}}$, $\langle \mu | \hat{\rho}_1 | \mu \rangle = \frac{1}{1 + N_0}$ are calculated for states with real amplitudes.

To obtain (II.3.23), the following relations are used

$$
\langle \omega_0 | \hat{\rho}_i | \omega_0 \rangle = \frac{1 + \Delta}{2\Delta^2} \langle \mu_0 | \hat{\rho}_i | \mu_0 \rangle - \frac{\gamma}{\Delta^2} \langle \mu_0 | \hat{\rho}_i | \mu_1 \rangle + \frac{\gamma^2}{2\Delta^2(1 + \Delta)} \langle \mu_1 | \hat{\rho}_i | \mu_1 \rangle
$$

$$
\langle \omega_1 | \hat{\rho}_i | \omega_1 \rangle = \frac{1 - \Delta}{2\Delta^2} \langle \mu_0 | \hat{\rho}_i | \mu_0 \rangle - \frac{\gamma}{\Delta^2} \langle \mu_0 | \hat{\rho}_i | \mu_1 \rangle + \frac{\gamma^2}{2\Delta^2(1 - \Delta)} \langle \mu_1 | \hat{\rho}_i | \mu_1 \rangle
$$

where $\Delta^2 = 1 - \gamma^2$.

The cut–off rate is therefore easily derived

$$
R_0 = \log \frac{2}{1 + 2\sqrt{p_{00}(1 - p_{11})}} = \log \frac{2}{1 + \sqrt{1 - \beta^2}} \qquad \text{(II.3.24)}
$$

In order to compare with the bounds, we prove

$$(i) \quad \left| \int \sqrt{p_0(\mu)p_1(\mu)} d^2\mu \right| = e^{-\frac{S}{4N_0}}$$

$$(ii) \quad \mathrm{Tr}\left[\hat{\rho}_0\hat{\rho}_1\right] = \frac{1}{1+2N_0} e^{-\frac{S}{1+2N_0}}$$

$$(II.3.25a, b)$$

On the other hand, as shown in [28], when the *a priori* probability are taken uniform, the performance are suboptimum. Furthermore, the operators used to compute the R_0 are derived from the o.d.o. which is optimum only for detection criterion. It is then clear that the expression given by (II.3.24) is in fact a lower bound denoted R_0^l and calculated as $R_0^l = \log \frac{2}{1+\sqrt{1-p_{11}}}$.

Now, depending on the value of S, the upper bound is given either by (II.3.25a) or (II.3.25b) as will be seen in the following expressions.

$$R_0^u = \log\left(J_0^u\right) = \log\left(\frac{2}{1+\frac{1}{\sqrt{1+N_0}}e^{-\frac{S}{1+N_0}}}\right)$$

$$(II.3.26b, c)$$

$$R_0^v = \log\left(J_0^v\right) = \log\left(\frac{2}{1+e^{-\frac{S}{4N_0}}}\right)$$

These bounds are plotted in the next figure for $N_0 = 0.1$. It is noticed that

 - $S \lesssim 0.02$: $R_0^l \le R_0 \le R_0^v$
 - $S \ge 0.02$: $R_0^l \le R_0 \le R_0^u$
 - $N_0 \lesssim 0.02$: $R_0^l \le R_0 \le R_0^u$

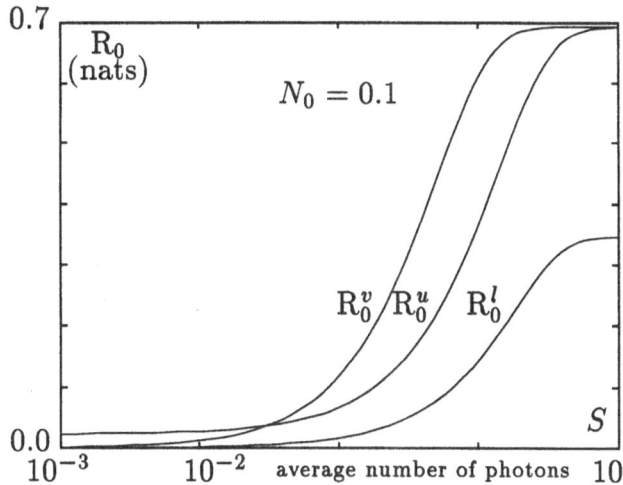

Fig. II.5.   Bounds for the cut–off rate versus the average number of coherent signal photons $S = |\mu|^2$ and for a noisy channel of a weak average number of photon thermal noise $N_0=0.1$.

II.3 References

[1]   R.J. Glauber: Phys. Rev. **131** 2766 (1963).

[2]   W.H. Louisell: *Radiation and Noise in Quantum Electronics* (Mc Graw Hill, New York 1964) Chapter VI, pp. 220–252.

[3]   R.G. Gallager: *Information Theory and Reliable Communication* (Wiley, New York 1965)

[4]   H. Takahashi: *Advances in Communication Systems* **1** (Academic Press 1966) 227–310.

[5]   J. Wozencraft, I. Jacobs: *Principles of Communication Engineering* (Wiley, New York 1968).

[6]   H.L. van Trees: *Detection, Estimation and Modulation Theory* (John Wiley & Sons, New York 1968). Part I.

[7]   A.V. Balakrishnan: *Communication Theory* (McGraw–Hill, New York 1968) Chapter V, pp. 192–215.

[8]   A. Spataru: *Théorie de la Transmission de l'Information* (Masson, Paris 1970)

[9]   R. Yoshitani: Jour. Stat. Phys. **2**, 347 (1970).

[10]   C.W. Helstrom, J.W.S. Liu, J.P. Gordon: Proc. IEEE **58** 1578 (1970).

[11]   T. Berger: *Rate distortion theory*, (Prentice Hall, Inc., Englewood Cliffs, N.J. 1971).

[12]   A.S. Holevo: Problemy Peredaci Informacii **9** 31 (1973).

[13]   R. S. Kennedy: Res. Lab. Electron., M.I.T., Cambridge, Quart. Prog. Rep. **108** pp. 219–225 (1973).

[14]   A.S. Holevo: Problems of Information Transmission **9** 117 (1975).

[15]   H. P. Yuen, R. S. Kennedy, M. Lax: I.E.E.E. **IT–21**125 (1975)

[16]   C.W. Helstrom: *Quantum Detection and Estimation Theory* (Academic Press, New York 1976).

[17]   E.B. Davies: *Quantum theory of open systems*, (Academic Press 1976).

[18]   R.J. McEliece: *The Theory of Information and Coding* (Addison–Wesley Pub. Co., Reading (USA) 1977).

[19]   C. Bendjaballah: in *Coherence and Quantum Optics IV*, ed. by L. Mandel and E. Wolf (Plenum Press, New York 1978). p. 241.

[20]   E.B. Davies: I.E.E.E. **IT–24** 596 (1978).

[21]   J.R. Pierce: I.E.E.E. **COM–26** 1819 (1978).

[22]   J. H. Shapiro: I.E.E.E. Trans. Inform. Theory, **IT–26** 490 (1980).

[23]   R.J. McEliece: I.E.E.E. Trans. Inform. Theory, **IT–27** 393 (1981).

[24]   J.L. Massey: I.E.E.E., **COM–29** 1615 (1981).

[25]   A.S. Holevo: *Probabilistic and Statistical Aspects of Quantum Theory* (North–Holland, Amsterdam 1982).

[26]   M. Charbit: *Information performance for communication channel with thermal noise* (LSS Internal Report 1988).

[27]   W.H. Zurek: Phys. Rev. A **131** 4731 (1989).

[28]   C. Bendjaballah, M. Charbit: I.E.E.E. Trans. Inf. Theory, **IT–35** 1114 (1989).

[29]   R.M. Gray: *Entropy and Information Theory* (Springer–Verlag, Heidelberg 1990).

[30]   W.H. Zurek: *Complexity, entropy and physics of information*, ed. by W.H. Zurek, (Addison–Wesley, Redwood City, USA 1990), Vol. **VIII**.

[31]   K.R.W. Jones: Annals of Physics, **207** 140 (1991).

[32]   T. Cover, J.A. Thomas: *Elements of information theory*, (John Wiley & Sons, New York 1991).

[33]   A. Vourdas: Phys. Rev. A, **46** 442 (1992).

[34]   M.J.W. Hall, M.J. O'Rourke : Eur. Opt. Soc. *preprint*.

[35]   T. Uyematsu, C. Bendjaballah : I.E.E.E. Trans. Comm. *to appear April 1995.*

[36]   J.M. Leroy, C. Bendjaballah : *in preparation.*

III. Direct Detection Processing

The method called direct–detection and its practical importance have been briefly analyzed in the Introduction from a sensitivity point of view. As seen in Sect. II, it is only suboptimum, as regards to the quantum detection performances which have been rigorously defined in [35],[37],[39],[58]. However, in many practical cases [38], it achieves interesting performances.

These receivers utilize the detectors e.g. photomultiplier tubes, avalanche diodes, p.i.n, $\cdots$ in order to measure directly the time–integrated low–power optical signal. In the treatments of the direct detection of the quantum limited signal, it is proved that the observations of the photodetector output constitute a *random point process* denoted "r.p.p.".

This r.p.p., analysed from the quantum mechanics [3], quantum stochastics [24], and classical [22] points of view, is of the doubly stochastic Poisson process type: conditionally to the given intensity (definition of the intensity as a random function of time for the point process treatment or as a stochastic operator for the quantum stochastic calculus, is always possible), the process is an inhomogeneous Poisson process for which there are many results. Mathematical proofs of these results are demonstrated using the calculus of probability in [34],[41],[43],[55],[70] to quote references that can be related to optical detection.

Hence, these results are very important because they induce the necessary relationship between the input and the output for the detection of light–field. Often the basic typical model is the pure Poisson counting process [5],[11],[12].

In optical communication, it is adequate for representing the detection process of a source of information of constant intensity.

. So far, the non–deterministic behaviour of the output process was supposed to be a result of the photocathode response. Now recent theories prove that the randomness is related to the source fluctuations (vacuum fluctuations) and the photodetector can be considered as being of a deterministic nature. The random fluctuations in the electron multiplication process is of no special importance for the present purposes.

In this section, we first recall the basic definitions and properties of the r.p.p. from the different points of view mentioned above. We then demonstrate the main results relevant to detection and information performances for direct detection.

III.1 Random Point Processes Theory

In the following, we present elements of the quantum and classical approaches to the characterization of the random point processes. Basic definitions and the main results will be briefly reviewed.

III.1.1 Quantum Formulation

Let us consider the electromagnetic field given by a collection of harmonic oscillators. The Hamiltonian is given by

$$\widehat{H} = \sum_l \hbar\omega_l \left(\widehat{a}_l^\dagger \widehat{a}_l + \widehat{a}_l \widehat{a}_l^\dagger\right) \qquad \text{(III.1.1)}$$

where the operators $\widehat{a}_l^\dagger$ and $\widehat{a}_l$ are

$$[\widehat{a}_{l_1}^\dagger, \widehat{a}_{l_2}] = \delta_{l_1,l_2} \qquad \text{(III.1.2a)}$$

$$[\hat{a}_{l_1}, \hat{a}_{l_2}] = [\hat{a}_{l_1}^\dagger, \hat{a}_{l_2}]^\dagger = 0 \qquad \text{(III.1.2}b\text{)}$$

The solution of the Schrödinger equation for the electrical field operator can be written (initial time $\vartheta_0 = 0$)

$$\hat{E}(\overrightarrow{r}, \vartheta) = i \sum_l \sqrt{\frac{\hbar \omega_l}{2\epsilon_0 V}} \overrightarrow{\epsilon_l} \left\{ \hat{a}_l e^{i(\overrightarrow{k_l} \overrightarrow{r} - \omega_l \vartheta)} - \hat{a}_l^\dagger e^{-i(\overrightarrow{k_l} \overrightarrow{r} - \omega_l \vartheta)} \right\}$$

$$\text{(III.1.2}c\text{)}$$

where V, the volume, is introduced in order to handle the discrete sum. To simplify the presentation, we will restrict our study to a single mode light field. It is convenient to write the electric field operator in the form [2,31]

$$\hat{E}(\overrightarrow{r}, \vartheta) = \hat{E}(\overrightarrow{r}, \vartheta)^+ + \hat{E}(\overrightarrow{r}, \vartheta)^- \qquad \text{(III.1.3)}$$

The correspondence between the classical and the quantum formulations, as established by Glauber [2] for a given density operator of the source $\hat{\rho} = |\{\alpha_l\}\rangle\langle\{\alpha_l\}|$, reads

$$\text{Tr}\left(\hat{\rho}\hat{E}\right) = E_{cl}$$

where $E_{cl} = 2i \, \Re e \sqrt{\frac{\hbar\Omega}{2V}} \alpha_l \, e^{i(\overrightarrow{k}_l \overrightarrow{r} - \omega_l \vartheta)}$. A simple model of photodetection consists in considering only the dipolar interaction $-\hat{P}\hat{E}\left(\overrightarrow{r}, \vartheta\right)$ for a given polarization between an atom of the photodetector and the electric field. The calculations to obtain the probability of detection of a photon are briefly summarized as follows [2,3].

We start with the matrix elements of the $U(\vartheta, \vartheta_0)$ time operator evolution between the initial $|\phi_g, \psi_i\rangle$ and final $|\phi_e, \psi_f\rangle$ states of the system (atom+field): $\langle\phi_e, \psi_f|U(\vartheta, \vartheta_0)|\phi_g, \psi_i\rangle$.

Let ω_{eg} be the frequency of the atomic transition $|\phi_g\rangle \rightarrow |\phi_e\rangle$ and $|\psi_i\rangle \rightarrow |\psi_f\rangle$ be the field transition.

The probability of the atomic transition at ϑ, the field being in the initial state defined by the density operator $\hat{R}$, is given by

$$P(\vartheta) = \mathrm{Tr}\left[\hat{R} \sum_{\{|\psi_f\rangle\}} \left| \langle \phi_e, \psi_f | U(\vartheta, \vartheta_0) | \phi_g, \psi_i \rangle \right|^2 \right]$$

$$= \frac{1}{\hbar^2} \int \int_{\Theta} d\vartheta_1 d\vartheta_2 \mathcal{F}(\omega_{eg}, (\vartheta_2 - \vartheta_1)) \times$$

$$\mathrm{Tr}\left[\hat{R}\hat{E}(\vec{r},\vartheta_1)^- \hat{E}(\vec{r},\vartheta_2)^+ \right]$$

$$(\mathrm{III.1.4})$$

where $\mathcal{F}(\omega_{eg}, (\vartheta_2 - \vartheta_1)) = \exp(i\omega_{eg}(\vartheta_2 - \vartheta_1))|\sigma_{eg}|^2$.

Using the expansion (III.1.3), only terms of $\hat{E}(\vec{r},\vartheta)^+$ will contribute in (III.1.4), because they conserve energy in the ionization process. The probability density to register a photoelectron in the time interval $\Theta = [\vartheta_0, \vartheta]$ is therefore given by

$$P[dN(\vartheta) = 1] = \int_{\vartheta_0}^{\vartheta} \int_{\vartheta_0}^{\vartheta} d\vartheta_1 d\vartheta_2 \mathcal{T}(\vartheta_1 - \vartheta_2) \times$$

$$\mathrm{Tr}\left[\hat{R}\,\hat{E}(\vec{r},\vartheta_1)^- \hat{E}(\vec{r},\vartheta_1)^+ \right]$$

$$= \eta\, \mathrm{Tr}\left[\hat{R}\,\hat{E}(\vec{r},\vartheta)^- \hat{E}(\vec{r},\vartheta)^+ \right]$$

$$(\mathrm{III.1.5})$$

where $\mathcal{T}(\vartheta)$ is the Fourier transform of the spectral response of the photodetector and η is a coefficient characterizing the detector.

III.1.1a Time-Occurence Probability Density

For a generalization of (III.1.5), let us consider n time intervals along the time axis and define the event $\mathcal{F}_n$ for which one photon is measured in $\{\vartheta_1, \vartheta_1 + d\vartheta_1\}$, another in $\{\vartheta_2, \vartheta_1 + d\vartheta_2\}, \cdots$, and any other in $\{\vartheta_n, \vartheta_n + d\vartheta_n\}$. The probability density of the event $\mathcal{F}_n$ is called time–occurence probability density and is given by

$$\text{Prob}\left\{[dN(\vartheta_1) = 1][dN(\vartheta_2) = 1] \cdots [dN(\vartheta_n) = 1]\right\}$$

$$= \text{Tr}\left\{\hat{R}\hat{\mathcal{O}}\left[\widehat{\mathcal{M}}(\vartheta_1)\widehat{\mathcal{M}}(\vartheta_2)\cdots\widehat{\mathcal{M}}(\vartheta_n)\right]\right\} \quad \text{(III.1.6}a\text{)}$$

where $\hat{\mathcal{O}}$ is the normal ordering operator. From (III.1.5), the operator $\widehat{\mathcal{M}}$ is

$$\widehat{\mathcal{M}}(\vartheta_i) = \sum_{\mu_1, \mu_2}\int\int d\overrightarrow{r_1}d\overrightarrow{r_2}\mathcal{D}_{\mu_1,\mu_2}\left(\overrightarrow{r_1}, \overrightarrow{r_2}\right) \times$$

$$\widehat{E}_{\mu_1}(r_1, \vartheta)^+\,\widehat{E}_{\mu_2}(r_2, \vartheta)^- \quad \text{(III.1.6}b\text{)}$$

where μ_1 and μ_2 are the polarizations of the potential vector and $\mathcal{D}_{\mu_1,\mu_2}\left(\overrightarrow{r_1}, \overrightarrow{r_2}\right)$ is the detector response.

III.1.1b Counting Probability

In the following we omit the spatial dependence on $\overrightarrow{r}$ and restrict ourselves to time properties of the field of a given direction and polarization. Here we define the basic operators and functions needed for the counting statistics. First, the counting operator is defined by

$$\hat{N}_T = \int_{\vartheta}^{\vartheta+T} d\zeta\,\widehat{\mathcal{M}}(\zeta) \quad \text{(III.1.7}a\text{)}$$

Therefore, from the previous equations (III.1.5)–(III.1.6), we may set

$$\text{Prob}\,[n; \vartheta, \vartheta + T] = \text{Tr}\left[\hat{R}\hat{O}\left\{\frac{\hat{N}_T^n}{n!}\exp\left(-\hat{N}_T\right)\right\}\right] \qquad \text{(III.1.7b)}$$

For a time independent, single mode–free field with $\widehat{M} = \eta \mathbf{a}^\dagger \hat{a}$ the generating function takes the form

$$G\,(s) = \text{Tr}\left[\hat{R}\hat{O}\left\{\exp\left(-s\eta T \hat{a}^\dagger \hat{a}\right)\right\}\right] \qquad \text{(III.1.8a)}$$

For this specific case, the coincidence probability is defined by

$$\text{Prob}\,\{[dN\,(\vartheta_1) = 1]\,[dN\,(\vartheta_2) = 1]\,...\,[dN\,(\vartheta_n) = 1]\}$$
$$= (\eta T)^n\,\text{Tr}\left[\hat{R}\hat{O}\left\{(\hat{a}^\dagger \hat{a})^n\right\}\right] \qquad \text{(III.1.8b)}$$

and from (III.1.7b), the counting probability density is given by

$$\text{Prob}\,[N_T = n] = \text{Tr}\left[\hat{R}\hat{O}\left\{\frac{(\eta T \hat{a}^\dagger \hat{a})^n}{n!}\exp\left(-\eta T \hat{a}^\dagger \hat{a}\right)\right\}\right] \quad \text{(III.1.8c)}$$

For a stationary field $[\hat{H}, \hat{R}] = 0$, and setting $\hat{R} = \sum_n R_{nn}|n\rangle\langle n|$ we obtain

$$\text{Prob}\left\{[dN\,(\vartheta_1) = 1]\,[dN\,(\vartheta_2) = 1]\,...\,[dN\,(\vartheta_n) = 1]\right\}$$
$$= \sum_{m=n}^{\infty} \eta^m \frac{m!}{(m-n)!}\,R_{mm} \qquad \text{(III.1.9)}$$

$$\text{Prob}\left[N_T = n\right] = (\eta T)^n \sum_{m=n}^{\infty} \binom{m}{n}(1 - \eta T)^{m-n}\,R_{mm} \quad \text{(III.1.10a)}$$

Remark: It should be emphasized that the number of photons absorbed by the detector cannot be larger than the total number of photons present in the field. This question has been studied by Mollow [16]. We refer the interested reader to the work of Davies [24] and Srinivas and Davies [54] for further details. Instead of (III.1.10a), the photoncounting distribution for a single mode–free field will be given by

$$p(n; T) = \left[1 - e^{-\eta T}\right]^n \sum_{m=n}^{\infty} \binom{m}{n} e^{-\eta T(m-n)} R_{mm} \qquad (\text{III.1.10}b)$$

from which it is easy to prove that $E[n] \leq \text{Tr}\left[\widehat{R}\widehat{a}^\dagger\widehat{a}\right]$ as required. Notice, however, that in most cases this correction is not significant.

In [64], it is demonstrated by solving the equation for the time evolution of the annihilation operator $\widehat{a}$ in the Heisenberg representation, that the probability of detecting a photon is of a Poisson type (III.1.8c).

III.1.2 Quantum Stochastic Formulation

Let, at the initial time s, the state of the field on the space denoted $\mathcal{K}$, be given by a coherent state $|\alpha\rangle$, $\widehat{\sigma} = |\alpha\rangle\langle\alpha|$. The field operators are the usual creation, annihilation and number $\widehat{a}^\dagger(s), \widehat{a}(s), \widehat{N}_s$ as already defined. According to Barchielli and col. (see [73] and references therein), adapted process operators (after interaction) are defined by integrals of these operators

$$\widehat{A}(t) \stackrel{\text{def}}{=} \int_0^t \widehat{a}(s)ds$$

The rules of the quantum stochastic calculus established by Hudson and col., are summarized in [73]. We just recall here that the derivative rule is $d\big(N_t \cdot \widehat{A}^\dagger(t)\big) = dN_t \cdot \widehat{A}^\dagger(t) + N_t \cdot d\widehat{A}^\dagger(t) + d\widehat{A}^\dagger(t)$.

Under the self–adjoint Hamiltonian $\widehat{H}$ in a separable Hilbert space $\mathcal{H}$, the system (defined by the density operator $\widehat{\rho}$) in interaction with the field, evolves with a unitary operator $\widehat{U}(t) \in \mathcal{H} \otimes \mathcal{K}$, given by the *stochastic Schrödinger equation*

$$\begin{cases} d\widehat{U}(t) = \Big[- \widehat{R}^\dagger \cdot d\widehat{A}(t) + \widehat{R} \cdot d\widehat{A}^\dagger(t) - \frac{1}{2}\widehat{R}^\dagger \cdot \widehat{R}\,dt - i\widehat{H}\,dt \Big]\widehat{U}(t) \\ \widehat{U}(0) = \mathbb{1} \end{cases}$$

where $\widehat{R}$ is the system–bounded operator on $\mathcal{H}$, which is of interest here. Thus, at time t, the density operator of the system can be written as

$$\widehat{\rho}(t) = \operatorname{Tr}\,{}_\mathcal{K}\Big[\widehat{U}(t)(\widehat{\rho} \otimes \widehat{\sigma})\widehat{U}(t)^\dagger\Big]$$

Let us now consider the device we have already called the photoelectron counter, which realizes the measurement of the observables, number and intensity operators, as follows

$$\begin{cases} \widehat{N}_{t,t\geq 0} = \widehat{U}(t)^\dagger\Big(\displaystyle\int_0^t \widehat{a}^\dagger(s)\widehat{a}(s)ds\Big)\widehat{U}(t) \\ \widehat{I}(t) = \displaystyle\int_0^t d\widehat{N}_s \end{cases} \qquad\qquad (\mathrm{III.1.11}a,b)$$

We introduce the characteristic operator (analogue of the characteristic functions (III.1.8a) and (III.1.14a) below)

$$\widehat{\phi}_t(u) \overset{\text{def}}{=} \exp\Big(iu\int_0^t d\widehat{N}_s\Big)$$

that obeys

$$\begin{cases} d\widehat{\phi}_t(u) = \widehat{\phi}_t(u)\left[\exp\left(iu-1\right)d\widehat{N}_t\right] \\ \widehat{\phi}_0(u) = \mathbb{1} \end{cases}$$

More general definition uses a time function $k(s)$ instead of a constant u. Therefore the characteristic function of the point process is given by

$$\Phi_t(u) \overset{\text{def}}{=} \text{Tr}_{\mathcal{H}\otimes\mathcal{K}}\left[\widehat{\phi}_t(u)\widehat{\rho}\otimes\widehat{\sigma}\right]$$

which yields the probability distribution of a compound Poisson type.

In the following, we will be working with a classical approach which is, in our opinion, more appropriate to interpret the questions of physics.

III.1.3 Classical Formulation

The $\mathcal{E}(\vartheta)$ incident is a quasi–monochromatic light field which is considered as a real, random, stationary function of time [4,15,23]. Its Fourier transform is written in the form $e(\omega) = a(\omega)e^{i\Phi(\omega)}$. The complex function of time $\mathcal{Z}(\vartheta)$, called the *analytic signal* is obtained from the electric field by the transforms

$$\mathcal{Z}(\vartheta) = \int_0^\infty 2a(\omega)\exp\left(i\left(\omega_0\vartheta + \Phi(\omega)\right)\right)d\omega \qquad \text{(III.1.12a)}$$

$$\mathcal{E}(\vartheta) = \Re e\{\mathcal{Z}(\vartheta)\} \qquad \text{(III.1.12b)}$$

the $\Re e\{\mathcal{Z}(\vartheta)\}$ and $\Im m\{\mathcal{Z}(\vartheta)\}$ being related Hilbert transforms.

In (III.1.12a), ω_0 and $\Phi(\omega)$ are the central frequency, and the phase shift, respectively. We denote as quasi–monochromatic a light field for which the bandwidth $\Delta\omega$ is such that $\Delta\omega \ll \omega_0$.

The emerging current intensity from the photodetector will be given by $J(\vartheta) = |\mathcal{Z}(\vartheta)|^2$. Although the quantum definition of the intensity operator is possible as $\hat{J} = \hat{O}\{\hat{E}^+\hat{E}^-\}$ or as (III.1.11b), the classical definition requires positivity of the intensity which is difficult to verify for some cases of the squeezed light.

Let $\gamma_J(\vartheta_1, \vartheta_2)$ be the normalized second order time correlation function of the random function $J(\vartheta)$. The coherence time of the stationary random function $J(\vartheta)$ is given by $\tau_c = 2 \int_0^\infty |\gamma(u)| du$ where $u = |\vartheta_1 - \vartheta_2|$.

Similarly, for the random function of space, we call the coherence length $r_c = 2 \int_0^\infty \left|\gamma_J\left(\vec{r}|u_0\right)\right| d\vec{r}$ where $\vec{r} = |\vec{r_1} - \vec{r_2}|$ and $u_0 = |\vartheta_1 - \vartheta_2|$.

Now for a fixed position of the 2D–photodetector of a spatial area A which is considered as to be of a constant gain, infinite bandwidth and without internal background noise, the instantaneous intensity becomes $\rho(\vartheta) = \eta J(\vartheta)$ with $\eta = \frac{\alpha A}{\hbar \omega_0}$ and η the *quantum efficiency* of the detector. In the following, the spatial properties which are especially important for image processing or space localization of parameters [32,33,74] are not considered. In this section, we set $\eta = 1$.

III.1.3a Main Properties of a Doubly Stochastic Poisson Process

It is useful to briefly recall the definitions and properties of the pure Poisson r.p.p. denoted *"P–r.p.p."* as follows

(i) Let $(\{\vartheta\}_n; n \geq 1)$ be a set of instants on the time axis. Let $\{\tau_i = \vartheta_i - \vartheta_{i-1}\}_{i=1}^n$ be independent random variables identically distributed (i.i.d.) with probability density $p(\tau_i = x) = \rho \, e^{-\rho x}$, for $x > 0$, ρ being deterministic, constant and > 0.

(ii) Let $T \in \mathbb{R}^+$ and N_T be the number of points occuring in T, defined such that $N_T = \sum_{n \geq 1} \delta(T - \vartheta_n)$. The $\{N_T\}$ is the counting random function.

(iii) Let $(\{\vartheta\}_n; n \geq 1)$ be the set of parameters $\rho > 0$ defined in (i). For $0 < T_1 < T_2 < \cdots < T_n$, the random increments $N_{T_1}, N_{T_2} - N_{T_1}, \cdots, N_{T_n} - N_{T_{n-1}}$ are independent. Moreover, for $T' < T$, $n \in \mathbb{N}$,

$$\text{Prob}\,[(N_T - N_{T'}) = n] = \frac{[\rho(T-T')]^n}{n!} \, e^{-\rho(T-T')}.$$

It is simple to demonstrate the reciprocal proposition.

(iv) Let $W_n = \sum_{i=1}^n \tau_i$. From the above definitions, $N_T \leq n$ iif $W_{n+1} \geq T$, $T > 0$, $n \in \mathbb{N}$, so that the equality $\text{Prob}[N_T < n] = \text{Prob}[W_n > T]$ is the relationship between counting and interval distributions.

We begin by establishing the main elements for the generalization of the above definitions to *conditional* "*P–r.p.p.*" [22,27,34].
We consider that the output of a photodetector receiving an optical signal of very weak power, is a set of narrow pulses over the time axis. These pulses correspond to the instantaneous emission of a

photon consequent to the absorption of a photon. This sequence of time instants $\{\vartheta_i\}$ constitutes a r.p.p..

Now, let this r.p.p. be characterized by the counting function that is the number of $\{\vartheta_i\}$ in the time interval T: $N(T) = \int_T dN(\vartheta)$, where $dN(\vartheta)$ are the random increments of $N(T)$. It can be seen from the theory of probability that the complete characterization of $N(T)$ requires the knowledge of the n–dimensional probability distribution for each n: $\text{Prob}[dN(\vartheta_1), ..., dN(\vartheta_n)]$ for every set $\{\vartheta_i\}_{i=1}^n$. Then, the following properties characterize the process

(i) $\forall i, \{\vartheta_i\} \equiv \vartheta$ $\qquad\qquad$ $\text{Prob}[dN(\vartheta)] > 1 = o(d\vartheta)$

$\qquad$ that is, the random variable $dN(\vartheta)$ can take only two values

$\qquad$ 0 or 1. Therefore, $\forall k, dN(\vartheta)^k = dN(\vartheta)$.

(ii) for different $\{\vartheta_i\}$

$$\text{Prob}\Big\{ [dN(\vartheta_1) = 1][dN(\vartheta_2) = 1] ... [dN(\vartheta_n) = 1] | \rho(\vartheta) \Big\}$$

$$= \rho(\vartheta_1)\rho(\vartheta_2)...\rho(\vartheta_n)\, d\vartheta_1 d\vartheta_2...d\vartheta_n$$

This is the classical form of counting probability density (III.1.8c). For photon counting purposes, let us deal with

$$\text{Prob}\Big[N(T) = \int_T dN(\vartheta) = n\Big]$$

To evaluate this integral, it is useful to start with the characteristic function

$$\phi_N(u|\rho) \overset{\text{def}}{=} E_N\Big[\exp\{iuN(T)\}\Big] = \exp\Big(iu\int_T dN(\vartheta)\Big)$$

The interval of length T is conveniently divided into equal subintervals δ_j such that

$$
\begin{cases}
\Delta N\left(\vartheta_j\right)=1 & \text{with the probability} \quad \rho\left(\vartheta_j\right)\delta_j \\[2mm]
\Delta N\left(\vartheta_j\right)=0 & \text{with the probability} \quad 1-\rho\left(\vartheta_j\right)\delta_j
\end{cases}
\qquad \text{(III.1.13)}
$$

Hence $\phi_N\left(u|\rho\right)=\exp\left(iu\sum_j \Delta N\left(\vartheta_j\right)\right)$ which becomes, taking into account (III.1.13)

$$
\begin{aligned}
\phi_N\left(u|\rho\right) &= \prod_j\left\{1+\left(\exp\left(iu\right)-1\right)\rho\left(\vartheta_j\right)\delta_j\right\} \\[2mm]
&\simeq \prod_j\left\{\exp\left(\left\{\exp\left(iu\right)-1\right\}\rho\left(\vartheta_j\right)\delta_j\right)\right\} \\[2mm]
&= \exp\left[\sum_j\left(\exp\left(iu\right)-1\right)\rho\left(\vartheta_j\right)\delta_j\right] \\[2mm]
&\underset{\delta_j\to 0}{\longrightarrow} \exp\left[\left(\exp\left(iu\right)-1\right)\int_T \rho\left(\vartheta\right)d\vartheta\right]
\end{aligned}
$$

Finally

$$
\begin{aligned}
\phi_N(u) &= \mathrm{E}_\rho\left[\exp\left(\left\{\exp\left(iu\right)-1\right\}\int_T \rho\left(\vartheta\right)d\vartheta\right)\right] \\[2mm]
p\left(n;T\right) &\equiv \mathrm{Prob}\left\{N\left(T\right)=n\right\} \\[2mm]
&= \frac{1}{2\pi}\int_{-\infty}^{\infty}\phi_N\left(u\right)\exp\left(-iuN\left(T\right)\right)du \\[2mm]
&= \mathrm{E}_\rho\left[\frac{\left(\int_T \rho\left(\vartheta\right)d\vartheta\right)^n}{n!}\exp\left(-\int_T \rho\left(\vartheta\right)d\vartheta\right)\right]
\end{aligned}
\qquad \text{(III.1.14a,b)}
$$

In view of (III.1.14a,b) and (III.1.7b), we can conclude that these approaches, briefly presented above, are "equivalent" at least for the one fold statistics.

Remark: In order to take into account the decay of the field state after interaction with the detector, it is sufficient to replace $\exp\left(-\int_T \rho(\vartheta)\,d\vartheta\right)$ in Eqs. (III.1.14a,b) by its first-order expansion $1 - \int_T \rho(\vartheta)\,d\vartheta$.

III.1.3b Photon Counting and Time Intervals Distributions

For the rest of this section, we will adopt this formulation which is more suitable for comparison with the experiment. In fact, treating the detector separately from the source makes the interpretation of the optical detection easier. The distribution given by (III.1.14b) is called counting probability density. For an inhomogeneous Poisson process, $\rho(\vartheta)$ is deterministic. For a pure Poisson process, $\rho(\vartheta)$ is constant. In both cases there is no need for averaging over the density of the process and (III.1.14b) takes a simple form.

As pointed out previously, instead of using only the counting process $\{N(t)\,|t \in [0,T]\}$ to characterize the occurrence properties of the r.p.p., it is sometimes useful to use the time–occurrence process $\{W(n;t)\,|n \in [1,\infty[\,\}$. In general, methods based on counting and time–occurrence distributions are quite different. This can easily be seen: the basic relation between these two distributions is given by (III.1.2b (iv)): $N(T) \leq n$ iff $W(n+1;t) \geq t$ $t >$ 0 ; $n \in [1,\infty[$. To calculate the needed distributions, it is convenient to start with the probability of the n–occurences at $\{\vartheta_i\}$ for each n

$$\text{Prob}\left(\vartheta_1, \vartheta_2, ..., \vartheta_n|\rho\right) = \rho(\vartheta_1)\,\rho(\vartheta_2)\,...\rho(\vartheta_n)\exp\left(-\int_{t_\alpha}^{t} \rho(\vartheta)\,d\vartheta\right)$$

$$(\text{III.1.15}a)$$

The marginal distribution is obtained by taking the average value over ρ

$$\text{Prob}\,(\vartheta_1, \vartheta_2, ..., \vartheta_n) = \mathrm{E}_\rho \left[\prod_{i=1}^{n} \rho\,(\vartheta_i) \exp \left(- \int_0^T \rho\,(\vartheta)\,d\vartheta \right) \right]$$

$$(\text{III}.1.15b)$$

The distribution given by (III.1.15b) is called time occurrence probability density. For $\rho\,(\vartheta_i) = \rho_k\,(\vartheta_i)$, we denote $p_k\,(\{\vartheta\}_i^n)$ the expression (III.1.15b).

The characteristic function and the counting probability distribution given by (III.1.14a, b) are the functions which are sufficient for the treatment of the r.p.p. in the context of this section. Other functions of interest can be defined as well. From

$$G_\rho\,(s; T) = \mathrm{E}_\rho \left[\exp \left(-s \int_0^T \rho\,(\vartheta)\,d\vartheta \right) \right] \qquad (\text{III}.1.15c)$$

and

$$G_N\,(s; T) = \mathrm{E}_\rho \left[\exp \left(\{\exp\,(-s) - 1\} \int_0^T \rho\,(\vartheta)\,d\vartheta \right) \right] \quad (\text{III}..1.15d)$$

which are the intensity and moment generating functions respectively, it is simple to derive the counting distribution

$$p\,(n; T) = \frac{(-1)^n}{n!} \frac{\partial^n}{\partial s^n} G_\rho\,(s; T) \bigg|_{s=1} \qquad (\text{III}.1.15e)$$

Because the origin of the time axis is arbitrary, this distribution is sometimes called *relaxed* [21].

Generalization to M-time joint counting is straightforward, at least formally. To obtain explicit formulas, some elementary algebra is, however, required.

For simplicity, we continue assuming stationarity of the light field

$$G_\rho(\{s_i\};\{T_i\}) = E_\rho\left[\exp\left(-\sum_{i=1}^{M} s_i \int_0^{T_i} \rho(\vartheta_i)\,d\vartheta_i\right)\right]$$

the M–fold counting distribution being given by

$$p(\{n_i\};\{T_i\}) = \left\{\prod_{i=1}^{M} \frac{(-1)^{n_i}}{n_i!}\frac{\partial^{n_i}}{\partial s_i^{n_i}}\right\}G_\rho(\{s_i\};\{T_i\})\Big|_{\{s_i\}=1}$$

Example:

Let us consider a simple case of the two fold counting [17]. For $M = 2$ and $t_1 = 0,\ T_1 = 0;\ t_2 = \tau;\ T_2 = T$,

$$r(n,T;\tau) = \frac{\text{Prob}\{[dN(0) = 1,[N(\tau,\tau+T) = n]\}}{\text{Prob}[dN(0) = 1]}$$

$$= \frac{1}{E[\rho]}E_\rho\left[\rho(0)\frac{\left[\int_\tau^{\tau+T}\rho(\vartheta)\,d\vartheta\right]^n}{n!}e^{-\int_\tau^{\tau+T}\rho(\vartheta)\,d\vartheta}\right]$$

A case of interest is $\tau = 0$,

$$r(n,T;0) \equiv q(n;T) = \frac{1}{E[\rho]}\frac{(-1)^n}{n!}\frac{\partial^n}{\partial s^n}\left\{\frac{1}{s}\frac{\partial}{\partial t}G_\rho(s;t)\right\}\Big|_{s=1,t=T}$$

Because the origin of the time measurement is a point of the r.p.p., this distribution can be called *triggered* [21].

The time intervals distributions can be calculated similarly. We can easily show

$$v_{n=1}(t) = E_\rho\left[\rho(t)\exp\left(-\int_\tau^{\tau+t}\rho(\vartheta)\,d\vartheta\right)\right] = -\frac{\partial}{\partial t}G_\rho(s;t)\Big|_{s=1}$$

$$(\text{III}.1.15f)$$

In the theory of the r.p.p. (see for example [11,12,34]) this is called the *waiting time* distribution. Similarly, one obtains

$$w_{n=1}(t; t_i = 0) = \frac{1}{\mathrm{E}[\rho]} \mathrm{E}_\rho \left[\rho(t_i = 0) \, \rho(t) \exp \left(- \int_\tau^{\tau+t} \rho(\vartheta) \, d\vartheta \right) \right]$$

$$= \frac{1}{\mathrm{E}[\rho]} \left\{ \frac{1}{s^2} \frac{\partial^2}{\partial t^2} G_\rho(s; t) \right\} \Big|_{s=1}$$

$$(\mathrm{III}.1.15g)$$

This is called the *life time* distribution. It is interesting to notice that for $\exp \left(- \int_\tau^{\tau+t} \rho(\vartheta) \, d\vartheta \right) \ll 1$, $w_{n=1}(t; t_i = 0) \sim \gamma_\mathrm{I}(t)$ where $\gamma_\mathrm{I}(t) = \mathrm{E}_\rho[\rho(0)\rho(t)]$ is the time correlation function of the light intensity. Generalization for higher n is shown below.

Let the moment–generating function written without taking into account the stationary properties of the light field

$$G_\rho(s; t_2, t_1) = \mathrm{E} \left[\exp \left(-s \int_{t_1}^{t_2} \rho(\vartheta) \, d\vartheta \right) \right]$$

We first make two partial differentiations with respect to t_1 and t_2.

$$\frac{\partial^2}{\partial t_1 \partial t_2} G_\rho(s; t_2, t_1) = -s^2 \mathrm{E} \left[\rho(t_1) \, \rho(t_2) \exp \left(-s \int_{t_1}^{t_2} \rho(\vartheta) \, d\vartheta \right) \right]$$

Then we take into account the stationary properties of the field, so that the functions only depend on $t = t_2 - t_1$. We arrive, for $n > 1$, at

$$\frac{\partial}{\partial t_2} G_\rho(s; t_2, t_1) = \frac{\partial}{\partial t} G_\rho(s; t); \quad \frac{\partial}{\partial t_1} G_\rho(s; t_2, t_1) = -\frac{\partial}{\partial t} G_\rho(s; t)$$

which lead to

$$w_n(t; t_i = 0) = \frac{1}{E[\rho]} \frac{1}{n!} E\left[\rho(t_i = 0) \left(\int_0^t \rho(\vartheta)\, d\vartheta\right)^{n+1} e^{-\int_0^t \rho(\vartheta)\, d\vartheta}\right]$$

$$= \frac{1}{E[\rho]} \frac{(-1)^n}{n!} \frac{\partial^n}{\partial s^n} \left\{\frac{1}{s^2} \frac{\partial^2}{\partial t^2} G_\rho(s; t)\right\}\Big|_{s=1}$$

Now we seek the relation between $w(t; t_i = 0|n) \equiv w(t; 0|n)$ and $p(n; t)$. We first notice that

$$\frac{\partial}{\partial t} p(n; t) = \frac{1}{(n-1)!} E\left[\rho(t) \left(\int_0^t \rho(\vartheta)\, d\vartheta\right)^{n-1} e^{-\int_0^t \rho(\vartheta)\, d\vartheta}\right]$$

$$- \frac{1}{(n)!} E\left[\rho(t) \left(\int_0^t \rho(\vartheta)\, d\vartheta\right)^n e^{-\int_0^t \rho(\vartheta)\, d\vartheta}\right]$$

A second derivative with respect to time shows

$$\frac{\partial^2}{\partial t^2} p(n; t) = E[\rho]\left\{w_n(t; 0) - 2w_{n-1}(t; 0) + w_{n-2}(t; 0)\right\}$$

Hence

$$\sum_{l=0}^m \frac{\partial^2}{\partial t^2} p(l; t) = E[\rho]\left\{w_m(t; 0) - w_{m-1}(t; 0)\right\}$$

so that

$$\begin{cases} w_n(t; 0) = \frac{1}{E[\rho]} \sum_{m=0}^n \sum_{l=0}^m \frac{\partial^2}{\partial t^2} p(l; t) \\ v_n(t) = -\sum_{l=0}^n \frac{\partial}{\partial t} p(l; t) \end{cases}$$
$$(\text{III.1.16}a, b)$$

III.1.3c Some Specific Cases

For counting processing, T is constant and $N(T) = n$ is the random variable. Here the number of counts equal to $0, 1, 2, ..., n$ is kept constant and the time interval $T = t$ is taken as a random variable. The following general expressions hold for a long t

$$G_\rho(s;t) = \sum_{n=0}^{\infty} s^n p(n;t) = \text{E}\left[\exp\left(-(1-s)\rho t\right)\right]$$

$$\overset{\text{def}}{=} g(s;t)\exp\left(-h(s;t)\right)$$

$$\frac{\partial}{\partial s}G_\rho(s;t)\bigg|_{s=0} = \text{E}\left[\rho t\, e^{-\rho t}\right] = p(n=1;t)$$

$$\frac{\partial}{\partial t}G_\rho(s;t)\bigg|_{s=0} = -v(t)$$

Similarly, one obtains $w(t) = \dfrac{2}{\text{E}[\rho]t^2}p(n=2;t)$.

(i) For coherent light, the r.p.p. is *memoryless*, and we easily find

$$p(n;t) \equiv r(n;\tau,t) \equiv q(n;t) = \frac{(\text{E}[\rho]t)^n}{n!}\exp\left(-\text{E}[\rho]t\right)$$

(ii) For thermal light, it will be shown (see III.2.3c) that the characteristic function $\phi_\rho(u;t) = G_\rho(-iu;T)$ can be obtained in the following form

$$\phi_\rho(u;t) = \text{E}[-iu\rho(t)] = \prod_m \frac{1}{1 - iu\text{E}[\rho]\lambda_m}$$

In the case of long coherence time, we have $\int_t \rho(\vartheta)d\vartheta \sim \text{E}[\rho]t$. Closed expressions can then be derived from $g(s;t) = \frac{1}{1+s\text{E}[\rho]t}$ and $h(s;t) = 0$, such that

$$p(n;t) = (1 - \nu_t)\nu_t^n \;, \quad q(n;t) = (n+1)\frac{(1-\nu_t)^2\,\nu_t^{n+1}}{\nu_t}$$

$$(\text{III.1.17}a,b)$$

where

$$\nu_t = \frac{\mathrm{E}[\rho]\,t}{1+\mathrm{E}[\rho]\,t}$$

$$(\text{III.1.17}c)$$

Moreover, for arbitrary τ, we obtain [21]

$$r(n,t;\tau) = \nu_t^n\left(1 + |\gamma_E(\tau)|^2\frac{n-\mathrm{E}[\rho]\,t}{1+\mathrm{E}[\rho]\,t}\right)$$

$$(\text{III.1.17}d)$$

Notice that the measurement of $r(n,t;\tau)$ is an indirect determination of the time correlation function of the light field. Recall that in the present case the normalized time correlation function of the light intensity is given by $|\gamma_I(\tau)| = 1 + |\gamma_E(\tau)|^2$.
The time interval distributions are simple to derive

$$v(t) = \frac{\mathrm{E}[\rho]}{(1+\mathrm{E}[\rho]t)^2} \;; \quad w(t) = \frac{2\mathrm{E}[\rho]}{(1+\mathrm{E}[\rho]t)^3}$$

$$(\text{III.1.18}a,b)$$

Notice that in order that the moments $\mathrm{E}[t^k]$, the calculation of the moments from these distributions requires limiting $t \in [0, \infty[$.

(iii) For a mixture of thermal of long coherence time and of photon average number $\mathrm{E}[\rho_t]$, and coherent radiation of photon average number $\mathrm{E}[\rho_c]$, the intensity distribution is

$$W(\rho) = \frac{m+1}{\mathrm{E}[\rho]}e^{-\left(m+\frac{(m+1)\rho}{\mathrm{E}[\rho]}\right)}I_0\left(\frac{2\sqrt{m(m+1)\rho}}{\mathrm{E}[\rho]}\right)$$

where $I_0(x)$ is the zero order Bessel function of imaginary argument. The parameter $m = \frac{\mathrm{E}[\rho_c]}{\mathrm{E}[\rho_t]}$ stands for the ratio of the single mode coherent intensity over the average thermal intensity and the average of the total intensity is $E[\rho] = E[\rho_c] + E[\rho_t]$.

The k–moment of the total intensity is found to be

$$E[\rho^k] = k! E[\rho]^k \frac{\exp(-\frac{m}{2})}{\sqrt{m}(m+1)^k} M_{\frac{k+1}{2},0}(m)$$

where $M_{\alpha,\beta}(m)$ is the Whittaker function. It is interesting to notice that the parameter $h = \frac{E[\rho^2]}{E[\rho]^2}$ is $\in [1,2]$.

The photon distribution can easily be computed

$$p(n;T) = \frac{(m+1)(E[\rho]T)^n}{(m+1+E[\rho]T)^{n+1}} e^{-\frac{mE[\rho]T}{m+1+E[\rho]T}} L_n\left(-\frac{m(m+1)}{m+1+E[\rho]T}\right)$$

$$(\text{III}.1.19)$$

where $L_n(z)$ is the Laguerre polynomial of degree n.

Another special light field derived from the previous cases concerns the superposition of coherent field and a Gaussian field of real or complex amplitude.

(i) For real amplitude field

$$W(\rho) = \frac{1}{\sqrt{2\pi \rho E[\rho]}} \exp\left(-\frac{\rho}{2E[\rho]}\right)$$

$$E[\rho^k] = (2k-1)!! E[\rho]^k$$

$$p(n;T) = \frac{\gamma(n+\frac{1}{2})}{\sqrt{\pi} n!} \frac{(2E[\rho]T)^n}{(1+2E[\rho]T)^{n+\frac{1}{2}}}$$

where $(2k-1)!! = 1.3.5...(2k-1)$ and $\gamma(x)$ is the factorial function, the parameter h being equal to 3.

(ii) For complex amplitude field specified by the real parameter λ such that for $\lambda = 0$, the resulting field is the superposition of

coherent with a thermal already studied. For $\lambda = 1$, the resulting field is the superposition of coherent with a Gaussian with real
amplitude just considered. To see this, use the asymptotic expansion $z \to \infty$: $I_0(z) \sim \frac{\exp(z)}{\sqrt{2\pi z}}$. We obtain

$$W(\rho) = \frac{1}{E[\rho]\sqrt{1 - \lambda^2}} e^{-\frac{\rho}{E[\rho](1 - \lambda^2)}} I_0\left(\frac{\lambda\rho}{E[\rho](1 - \lambda^2)}\right)$$

$$p(n; T) = \frac{\langle n \rangle^n \sqrt{1 - \lambda^2}}{((1 + \langle n \rangle)^2 - \lambda^2)^{n+1}} P_n\left(\frac{1 + \langle n \rangle}{\sqrt{(1 + \langle n \rangle)^2 - \lambda^2}}\right)$$

where $\langle n \rangle = E[\rho]T(1 - \lambda^2)$, and $P_n(x)$ is the n–th Legendre func-
tion of real argument. Here, $E[\rho^k] = k! E[\rho]^k (1 - \lambda^2)^{\frac{k}{2}} P_k(\frac{1}{\sqrt{1-\lambda^2}})$
which yields $\forall \lambda$, $h \in [2, 3]$. Details concerning the special func-
tions $I_0(x), L_n(z), M_{\alpha,\beta}(m), \gamma(x), P_n(x)$ can be found in [8].

For coherent squeezed light (μ, ν, S), the counting probability
density has already been calculated by Yuen [36] for pure input
coherent signal and by Helstrom [48] for a input coherent signal
corrupted by a thermal background in the case of real μ, ν and for
a suitable choice of the phase of the coherent input signal $(\theta = \frac{\pi}{2})$

$$p(n; t) = \frac{1 - v}{\sqrt{1 - w^2}} e^{-\frac{(1 - v)}{1 - w} S_c} \left(\frac{v - w^2}{1 - w^2}\right)^n$$

$$\sum_{m=0}^{n} \frac{n!}{(n - m)!} \left[\frac{(1 - v) w}{v - w^2}\right]^m [H_m(c)]^2 \qquad \text{(III.1.20a)}$$

In the case of no thermal background, this distribution takes a
simpler form using the conventional notations $v = w^2 = \tanh(r)$,
$S_c = S - \sinh^2(r), c^2 = \frac{e^{2r} S_c}{\sinh(2r)}$, $H_n(c)$ being the Hermite polynomials [8]

$$p(n) = \frac{1}{\cosh(r)} e^{-S_c\left(1+\tanh(r)\right)} \frac{\left(\frac{\tanh(r)}{2}\right)^n}{n!} [H_n(c)]^2 \qquad \text{(III.1.20b)}$$

from which we derive

$$\sigma_n^2 = S e^{-2r} + \frac{\sinh^2(2r)}{2} \qquad \text{(III.1.20c)}$$

that is plotted in the next figure for the three types of light we analyzed. In fact, we must take into account that r depends on S. With the notations $f = \frac{S_c}{S}$, we find $r = \mathrm{atanh}\left(\sqrt{\frac{(1-f)S}{1+(1-f)S}}\right)$. Clearly, another relation is needed to fix the value of f for given S. This is simply done by seeking the value $f_m \sim 1 - \frac{\sqrt{S}}{3} e^{-\sqrt{\frac{2S}{3}}}$ of f yielding the minimum of $P_e = \frac{1}{2}p(0;t)$. Thus, the variances for different light fields can be compared. This is done in following Fig. III.1b.

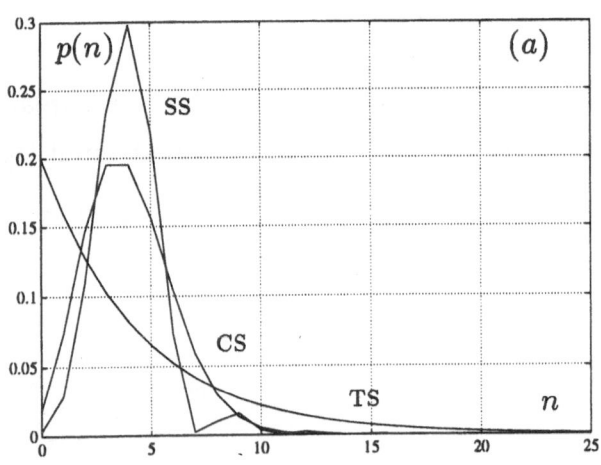

Fig. III.1a.   The curves display the photocount probability distributions for coherent squeezed state (SS) with $f = 0.8675$, coherent state (CS) and thermal state (TS) versus the photon number n. The received number of photons is $S = 4$.

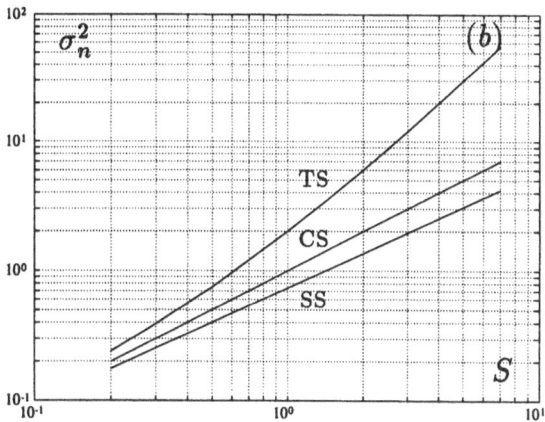

Fig. III.1b. The variances of these fields are plotted in Fig. (b) versus the received number of photons S. To plot (SS) the needed value of f for each value of S is taken such that the value of P_e is minimum.

It is interest to notice that the curve $f_m(S)$ exhibits a minimum for $f_c \sim 0.8535$ at $S_c \sim 1.5$. Now, let us denote

$$g(s;t) = \sqrt{\frac{1 - \Delta(t)}{1 - s^2 \Delta(t)}}, \quad \text{and} \quad h(s;t) = \lambda t \frac{1 + \sqrt{\Delta(t)}}{1 + s\sqrt{\Delta(t)}}(1 - s)$$

where we set $t\lambda = S_c, t\mu = (1 - f)S = S_n, \Delta(t) = \frac{t\mu}{1+t\mu}$. The time interval distributions are therefore given by [69,70]

$$v(t) = p(n = 1;t) \frac{\frac{g_t^{(1)}(0;t)}{g(0;t)} - h_t^{(1)}(0;t)}{h_s^{(1)}(0;t) - \frac{g_s^{(1)}(0;t)}{g(0;t)}} \qquad (\text{III.1.20}d)$$

$$w(t) = \frac{2p(n = 2;t)}{E[\rho]}$$

$$\frac{\frac{g_t^{(2)}(0;t)}{g(0;t)} h_s^{(1)}(0;t) - 2\frac{g_t^{(1)}(0;t)}{g(0;t)} h_t^{(1)}(0;t) - h_t^{(2)}(0;t) + \left(h_t^{(1)}(0;t)\right)^2}{\frac{g_s^{(2)}(0;t)}{g(0;t)} h_s^{(1)}(0;t) - 2\frac{g_s^{(1)}(0;t)}{g(0;t)} h_s^{(1)}(0;t) - h_s^{(2)}(0;t) + \left(h_s^{(1)}(0;t)\right)^2}$$

$$(\text{III.1.20}e)$$

where $g_t^{(l)}(0;t) = \frac{\partial^l}{\partial t^l} g(s;t)|_{s=0}$, $h_t^{(l)}(0;t) = \frac{\partial^l}{\partial t^l} h(s;t)|_{s=0}$ and

other similar notations for $g_s^{(l)}(0;t) = \frac{\partial^l}{\partial s^l} g(s;t)|_{s=0}$, $h_s^{(l)}(0;t) = \frac{\partial^l}{\partial s^l} h(s;t)|_{s=0}$. A typical example of the distributions (III.1.20d) and (III.1.20e) is displayed below in the next figure.

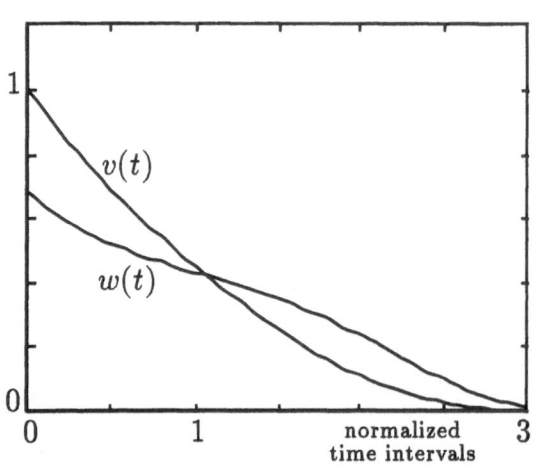

Fig. III.2. Time intervals distributions for coherent squeezed light: $v(t)$ waiting time; $w(t)$ life time, versus time intervals normalised to the average of the total number of photons, for $S = 1$ and $f = 0.875$.

The antibunching effect [59,61,69] is clearly observed for $t \leq 1$. In fact, from (III.1.15f, g) where $E[\rho] \equiv \langle \rho \rangle$), it is seen that $w(0) \leq v(0) \rightarrow \frac{\langle \rho^2 \rangle}{\langle \rho \rangle} \leq \langle \rho \rangle \Rightarrow \sigma_\rho^2 \leq 0$, from which it results that the interpretation of such a non–classical effect occuring in the "intensity" measurements, is inappropriate from the theory of compound random point processes.

III.1.3d Applications

As a first application of the results of this section, we want to derive $w(t_n)$ the probability distribution that $t_n = \sum_{i=1}^n \tau_i$ where $\tau_i = \vartheta_{i+1} - \vartheta_i;\ \forall i, \vartheta_i \neq 0$. Let us first calculate this distribution for a Poisson process. We may use the above expressions or derive it more simply knowing that in this case, the τ_i are i.i.d.. Thus

$$\phi_{t_n}(u) = e^{iut_n} = e^{iu\sum_{j=1}^{n}\tau_j} = \prod_{j=1}^{n} e^{iu\tau_j} = \left(\frac{\rho}{\rho - iu}\right)^n$$

where we set $E[\rho] = \rho$ and with $w(\tau_j) = \rho e^{-\rho\tau_j}$. Therefore,

$$w(t_n) = \frac{1}{2\pi}\int_{-\infty}^{\infty} \phi_{t_n}(u)\, e^{-iut_n}\, du = \frac{1}{2\pi}\int_{-\infty}^{\infty}\left(\frac{\rho}{\rho - iu}\right)^n e^{-iut_n}\, du$$

$$= \frac{(\rho t_n)^{n-1}}{(n-1)!}\, \rho\, e^{-\rho t_n}$$

As a generalisation of the previous results, we investigate the probability distribution of the sum of a finite number of correlated variables.

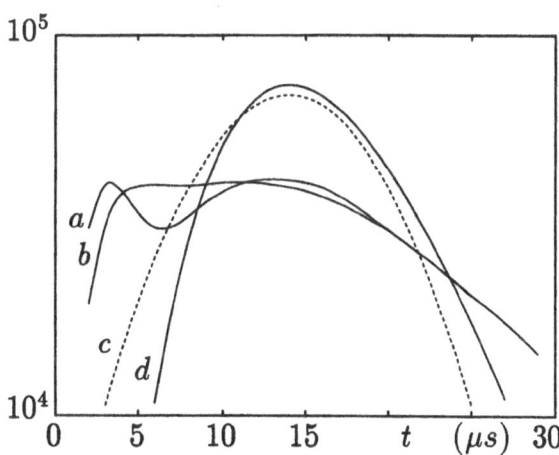

Fig. III.3. Probability distribution $w(t_n)$ of the sum of n intervals for $n = 7$; $\tau_c = 5.65\mu s$; $E[\rho] = 5.10^5 s^{-1}$. (a): Thermal light with a Gaussian profile. (b): Thermal light with an exponential profile. (c): Gaussian distribution $E[t] = \frac{n}{E[\rho]}$; $\sigma^2 = \frac{n+1}{E[\rho]^2}$. (d): Laser light (independent intervals), $w(t_n) = \frac{\rho^{n+1}}{n!}t^n \exp(-\rho t)$.

As before, let $\tau_i = \vartheta_{i+1} - \vartheta_i$ and $t_n = \sum_{i=0}^{n} \tau_i$. The typical curves of the probability distribution of t_n are displayed above. As previously seen, the "r.p.p.", being a compound Poisson process, the $\{\tau_i\}$ are correlated random variables: $\xi_{ij} = E[\tau_i \tau_j] \neq E[\tau_i]E[\tau_j]$ so that, changing the notations to read $w(t_n) \equiv w_n(t; 0)$, the distribution of the t_n for n quite high (in Fig. III.3, $n = 7$) is not Gaussian, as it should be for i.i.d..

As a matter of fact, let us be given a set of r.v. $\{x_k\}$ where $\langle x_k \rangle = \eta$, and $\sigma_{x_k}^2 = \sigma^2$ are their average and variance.

Let us consider the r.v. taking the values $y = \sum_{k=1}^{M} \frac{x_k - \eta}{\sigma\sqrt{M}}$ for fixed M and assume that the r.v. $\{x_k\}$ are i.i.d. . The central limit theorem states that the probability density $g(y)$ tends to the normal distribution $\mathcal{N}(0,1) = \frac{1}{\sqrt{2\pi}} \exp(-\frac{y^2}{2})$. The demonstration can be simplified such that, given $\Phi_x(u) = E\left[e^{iu\frac{x_k-\eta}{\sigma}}\right]$ which leads to

$$\Phi_x(\frac{u}{\sqrt{M}}) \sim 1 - \frac{u^2}{2M} + 0\left(\frac{u^3}{6M^{\frac{3}{2}}}\right) \sim \exp(-\frac{u^2}{2M})$$

we will have, for $M \to \infty$ and $\sigma_{x_k}^2 > \mu > 0$ and $\langle |x_k - \eta|^3 \rangle < \nu$ with $\nu > 0$,

$$\Phi_y(u) = E\left[e^{iuy}\right] = \prod_{k=1}^{M} \left(\Phi_x(\frac{u}{\sqrt{M}})\right) = \left(\Psi_x(\frac{u}{\sqrt{M}})\right)^M = e^{-\frac{u^2}{2}}$$

which is the characteristic function of a normal r:v.

In the case that M is random but independent of $\{x_k\}$, it is simple to prove that for the r.v. $z = \sum_{k=1}^{M} x_k$, we have $\langle z \rangle = \eta\langle M \rangle$ and $\langle z^2 \rangle = \eta^2\langle M^2 \rangle + \sigma^2\langle M \rangle$.

Finally, notice that the central limit theorem holds in certain situations even for not independent, not i.i.d. random variables.

The quantum analog of the central limit theorem has already been proved [15]. The demonstration is briefly reproduced in subsection I.2.8. Therefore, in connection with the application of the central limit theorem, it is of interest to study the deviation of $w(t_n)$ from a Gaussian whose distribution is, of course, reached for $n \to \infty$ [42]. This is shown in the previous figure where several curves $w(t_n)$ are displayed and compared for various types of light (coherent, thermal) and for different types of the second–order time correlation functions (Gaussian, exponential). The number of added intervals must be kept constant: $1 \ll n \ll \infty$. A detailed analysis of $w(t_n)$ versus n has been published [29].
We may briefly comment on the curves of Fig. III.3.

1. the curves (a) and (b) deviate significantly from the Gaussian curve (c)

2. the form of the spectral profile is important

3. for independent intervals (d), the curve approaches the Gaussian curve (c)

4. for (a) and (b) the behavior can be understood as follows: the added variables can be split in two sets

    – the first maximum is related to the distribution of the variables for which $\xi_{ij} \neq E[\tau_i]E[\tau_j]$. This occurs for short t.

    – the second maximum occurs around $E[t] = \frac{n}{E[\rho]}$ and can be related to the distribution of the variables where $\xi_{ij} = E[\tau_i]E[\tau_j]$.

5. for $n \to \infty$ most of the τ_i are i.i.d. and the distribution con-
centrates around the later maximum $\mathrm{E}[t] = \frac{n}{\mathrm{E}[\rho]}$.

More details are given in [28,29].

The correlation between intervals can indeed be studied from the coefficients ξ_{ij}. It is shown below how to calculate them [1].

Let us denote by $w_n\,(t|\vartheta_0 = 0) \equiv w\,(t_n)$ the probability density distribution, given that the time interval of length t has been detected, subject to the condition that $\vartheta_0 = 0$ and $n - 1$ points have occured. As shown in (III.1.16), we have a relation between $w_n\,(t|\vartheta_0 = 0)$ and $p\,(n;t)$.

We want to prove a relation between $\mathrm{E}[\tau_0 \tau_n]$ and $p\,(n;t)$. To do this, it is convenient to use the Laplace transforms of $p\,(n;t)$ and $w_n\,(t|\vartheta_0 = 0)$. Let

$$\mathcal{P}\,(n;\zeta) = \int_0^\infty p\,(n;t)\exp\,(-\zeta t)\,dt$$

and

$$\mathcal{W}\,(n;\zeta) = \int_0^\infty w_n\,(t|\vartheta_0 = 0)\exp\,(-\zeta t)\,dt = \mathrm{E}_t[\exp\,(-\zeta t)]$$

By expanding in series the function $\exp\,(-\zeta t)$ it is easy to verify that

$$\begin{cases} \left.\frac{\partial^k}{\partial \zeta^k}\mathcal{P}\,(n;\zeta)\right|_{\zeta=0} = (-1)^k \int_0^\infty t^k p\,(n;t)\,dt \\ \left.\frac{\partial^k}{\partial \zeta^k}\mathcal{W}\,(n+1;\zeta)\right|_{\zeta=0} = (-1)^k\ \mathrm{E}\left[\left(\sum_{i=0}^n \tau_i\right)^k\right] \end{cases}$$

One therefore obtains

$$\mathcal{P}\,(n;\zeta) = \frac{\mathrm{E}[\rho]}{\zeta^2}\{\mathcal{W}\,(n+1;\zeta) - 2\mathcal{W}\,(n;\zeta) + \mathcal{W}\,(n-1;\zeta)\}$$

Taking the k–th derivatives of both sides leads finally to:

$$\frac{\partial^k}{\partial \zeta^k} \mathcal{P}(n;\zeta)\Big|_{\zeta=0} = \sum_{l=0}^{k} (-1)^{k-l} \frac{k!}{l!} (k-l+1) \times$$

$$\left[\frac{1}{\zeta^{k-l+2}} \left\{ \frac{\partial^k}{\partial \zeta^k} \{W(n+1;\zeta) - 2W(n;\zeta) + W(n-1;\zeta)\} \right\}\Big|_{\zeta=0} \right]$$

For example, for $k = 0$,

$$\mathcal{P}(n;0) = \frac{1}{E[\rho]} \int_0^\infty p(n;t)\, dt$$

$$= \lim_{\zeta \to 0} \frac{1}{\zeta^2} \left\{ W(n+1;\zeta) - 2W(n;\zeta) + W(n-1;\zeta) \right\}$$

$$= \frac{1}{2} E\left[\left(\sum_{i=0}^{n} \tau_i \right)^2 - 2 \left(\sum_{i=0}^{n-1} \tau_i \right)^2 + \left(\sum_{i=0}^{n-2} \tau_i \right)^2 \right] = E[\tau_0 \tau_n]$$

III.2 Detection Performance

Because we are mainly interested in detection studies, questions concerning estimation of parameters via the r.p.p. observations are not discussed here. We refer the interested reader to [34,41,43,74].

III.2.1 Detection of Point Processes

The most general problem consists of deciding which of the M hypotheses $\{H_i\}_{i=1}^{M}$ governs the statistics of an observed path $\{y_\vartheta; 0 \le \vartheta \le T\}$. To simplify the presentation, let us consider a binary detection of an unknown signal in thermal background of constant intensity The constant value of ρ_0 in H_0 can be understood because the intensity of thermal noise with a very short correlation time (all the fluctuations are time–integrated), is constant. It can be shown that a rigorous solution to this problem requires the use of the causal minimum mean–square error.

As demonstrated, the quasi–optimum solution can be obtained by using a Kalman–Bucy filter (see for example [34]).

We emphasize that solving this decision problem yields a complete reconstruction of the required signal [19,20,26]. Unfortunately, implementing the decision by numerical computations appears very difficult to perform. However, for the binary decision problem, some results can be obtained for special situations by studying the likelihood ratio test (LRT) criterion [27,34].

For $k = 0, 1$, we easily obtain for an equal *a priori* probability for both hypotheses

(i) $H_0 : \lambda_0(\vartheta) = \rho_0$;

$$p_0\left(\{\vartheta_i\}_{i=1}^n\right) = \rho_0^n e^{-\rho_0 T}$$

(ii) $H_1 : \lambda_1(\vartheta) = \rho_S(\vartheta) + \rho_0$;

$$p_1\left(\{\vartheta_i\}_{i=1}^n\right) = \prod_{i=1}^n \left(\rho_S(\vartheta_i t) + \rho_0\right) e^{-\rho_0 T - \int_0^T \rho_S(\vartheta) d\vartheta}$$

Hence for the LRT:

$$\Lambda\left(\{\vartheta_i\}_{i=1}^n\right) = \frac{p_1\left(\{\vartheta_i\}_{i=1}^n\right)}{p_0\left(\{\vartheta_i\}_{i=1}^n\right)} \underset{H_0}{\overset{H_1}{\underset{<}{>}}} \mu \qquad (\text{III}.2.1a)$$

where

$$\Lambda\left(\{\vartheta_i\}_{i=1}^n\right) = E_{\rho_S}\left[\prod_{i=1}^n \left(1 + \frac{\rho_S(\vartheta_i)}{\rho_0}\right) \exp\left(-\int_0^T \rho_S(\vartheta) d\vartheta\right)\right]$$
$$(\text{III}.2.1b)$$

Notice that here n is a random variable. The expression of the LRT is expanded

$$\Lambda(\{\vartheta_i\}_{i=1}^n) \sim E_{\rho_S}\left[\exp\left(-\int_0^T \rho_S(\vartheta)\,d\vartheta\right)\right]$$

$$+\frac{1}{\rho_0}\sum_{i=1}^n E_{\rho_S}\left[\rho_S(\vartheta_i)\exp\left(-\int_0^T \rho_S(\vartheta)\,d\vartheta\right)\right]$$

$$+\frac{1}{\rho_0^2}\sum_{i,j=1}^n E_{\rho_S}\left[\rho_S(\vartheta_i)\rho_S(\vartheta_j)\exp\left(-\int_0^T \rho_S(\vartheta)\,d\vartheta\right)\right]$$

$$+\cdots$$

$$\text{(III.2.1c)}$$

Only the second term of (III.2.1c) depends on the observations. In the case of high level noise $\rho_0 \gg E[\rho_S]$, n being a random variable, the equation (III.2.1c) can be approximated by the first order term in $\frac{1}{\rho_0}$. Moreover, it can be shown that for a large class of stationary processes, this test is a linear function of the observations $\{\vartheta_i\}_{i=1}^n$. In fact, for a deterministic signal

$$L(\{\vartheta_i\}_{i=1}^n) = \log \Lambda(\{\vartheta_i\}_{i=1}^n)$$

$$= \sum_{i=1}^n -e^{-\int_0^T \rho_S(\vartheta)\,d\vartheta} + \log\left(1 + \frac{\rho_S(\vartheta_i)}{\rho_0}\right) \overset{H_1}{\underset{H_0}{\gtrless}} \gamma$$

$$\text{(III.2.1d)}$$

the decision is based on the function $d(\vartheta) = \log\left(1 + \frac{\rho_S(\vartheta)}{\rho_0}\right)$ which can be more precisely studied for small and large signal–to–noise ratios (SNR).

(i) SNR $\ll 1$: $d(\vartheta) \sim \frac{\rho_S(\vartheta)}{\rho_0}$, the solution to the decision problem is a matched filter receiver of constant impulse response in $[0,T]$.

(ii) SNR $\gg 1$: the density $\rho_S(\vartheta_i)$ can be considered as a random variable so that $\rho_S(\vartheta_i) = n_i\rho_{S,i}$. Hence $L = \sum_{i=1}^n n_i \log \frac{\rho_{S,i}}{\rho_0}$.

Therefore $\{n_i\}$ is a *sufficient statistics*. The solution to this decision problem is the photon counting receiver. It is analyzed in detail in the next section.

III.2.2 Photoncounting Processing :
Detection of Coherent Signal in Thermal Noise

The detection of a coherent signal in the presence of thermal noise of constant intensity has been extensively studied by Helstrom [37]. Here, we will just mention that closed analytical results for ROC can be obtained for thermal light of arbitrary time correlation. First for long coherence time and from (III.1.19), the LRT can be proved to verify

$$\begin{cases} \Lambda(\lambda) = \exp[-S(1 - \nu_t)]\mathrm{L}_\lambda\left(-\frac{S(1-\nu_t)^2}{\nu_t}\right) \\ \Lambda(\lambda + 1) = \frac{2\lambda + 1 + \frac{S(1-\nu_t)^2}{\nu_t}}{\lambda+1}\Lambda(\lambda) - \lambda\Lambda(\lambda - 1) \end{cases} \qquad (III.2.2a)$$

where $\mathrm{L}_n(x)$ is the Laguerre polynomial of degree n. We have

$$\begin{cases} \Lambda(1) = \left(1 + \frac{S(1-\nu_t)^2}{\nu_t}\right)\exp\left(-S(1 - \nu_t)\right) \\ \Lambda(0) < 1 \end{cases} \qquad (III.2.2b)$$

Therefore, $Q_d(\lambda + 1) = Q_d(\lambda) + \Lambda(\lambda)\left[Q_0(\lambda + 1) - Q_0(\lambda)\right]$.

Now for H_0 and from (II.1.16), $Q_0(\lambda + 1) = \nu_t Q_0(\lambda)$. Thus, because $\nu_t < 1$ we have $\Lambda(\lambda + 1) > \Lambda(\lambda) > 0$.

Remark: We notice that the curves ROC, i.e. Q_q versus Q_0 are non– decreasing functions and can be constructed segment by segment without any other computation.

For an arbitrary coherence time, general results can be obtained from numerical computations using the method recalled below.

The light field $\mathcal{E}(\vartheta)$ that is considered complex and Gaussian, can be expanded in an orthogonal basis on the interval $T = \left(-\frac{t}{2}, \frac{t}{2}\right)$: $\mathcal{E}(\vartheta) = \sum_m c_m \Phi_m(\vartheta)$ where Φ_m is a complete basis of the orthonormalized eigenfunctions associated with λ_m, the eigenvalues of the integral equation of the Freedholm type

$$\int_T \gamma_{\mathcal{E}}(\vartheta; \vartheta') \Phi_m(\vartheta')\, d\vartheta' = \lambda_m \Phi_m(\vartheta; t) \qquad (III.2.3a)$$

with $\int_T \Phi_m(\vartheta) \Phi_n^*(\vartheta)\, d\vartheta = \delta_{mn}$. The statistical properties of the field allow us to write

$$\begin{cases} E[c_m^* c_n] = \lambda_m \delta_{mn} \\ P(\{c_m\}) = \prod_m \frac{1}{\pi \lambda_m} \exp\left(-\frac{|c_m|^2}{\lambda_m}\right) \end{cases}$$

The integrated intensity over T is $\rho(T) = \int_T \mathcal{E}(\vartheta) \mathcal{E}^*(\vartheta) = \sum_j |c_j|^2$ and its generating function can be written in the form

$$G_t(s; T) = E\left[e^{-s\rho(T)}\right] = \int_T e^{-s \sum_j |c_j|^2} P(c_m)\, dc_m$$

$$= \prod_m \frac{1}{1 + sE[\rho]\lambda_m} \qquad (III.2.3b)$$

To calculate the eigenvalues $\{\lambda_m\}$, the solution to the integral equation (III.2.3a) is required.

This can be achieved for the $\gamma_{\mathcal{E}}(\vartheta; \vartheta') = \exp(-\Gamma|\vartheta - \vartheta'|)$ typical second order time correlation function, although in the case where the time correlation function of the light field differs from the exponential profile, some results can be obtained using the numerical method developed by Lachs [25]. The eigenfunctions verify the differential equation

$$\lambda_m \frac{\partial^2}{\partial t^2} \Phi_m\left(\vartheta;t\right) + \Gamma\left(2 - \lambda_m\Gamma\right)\Phi_m\left(\vartheta;t\right) = 0$$

We obtain

$$\lambda_m = \left(\Gamma T\right)^2 \left(\left(\Gamma T\right)^2 + \chi_m^2\right) \qquad \text{(III.2.3c)}$$

where χ_m are the roots of one of the following equations

$$\begin{cases} 2\chi \tan\chi = \Gamma T \\ 2\chi \cot\chi = -\Gamma T \end{cases} \qquad \text{(III.2.3d)}$$

so that only the positive roots are taken into account. Therefore, it is easy to calculate the needed distributions [13]

$$p\left(n;T\right) = \frac{1}{n!} \sum_{k=0}^{n-1} (-1)^{n+k+1} \binom{n}{k} k! p\left(k;T\right) D^{(n-k)}\left(s\right)\Big|_{s=1}$$

$$\text{(III.2.3e)}$$

where the following notations are set

$$D^{(l)}\left(s\right) = e^{-\Gamma T}\left(\frac{\Gamma T^2}{\zeta}\right)^l \times$$

$$\left[\frac{\Gamma T}{2} I_l\left(\zeta\right) + \zeta\left(1 + \frac{1}{2\Gamma T}\right) I_{l-1}\left(\zeta\right) + \frac{\zeta^2}{2\Gamma T} I_{l-2}\left(\zeta\right)\right] G_t\left(s;T\right)$$

$I_l\left(\zeta\right)$ are the l-order modified spherical Bessel functions of the first kind [8], and

$$G_t\left(s;T\right) = \frac{\exp\left(\Gamma T\right)}{\cosh\left(\zeta\right) + \mathcal{D}\left(\Gamma T, \zeta\right)\sinh\left(\zeta\right)}$$

where we set $\zeta = T\sqrt{\Gamma^2 + 2\Gamma s\mathrm{E}\left[\rho\right]}$ and $\mathcal{D}\left(\Gamma T, \zeta\right) = \frac{\Gamma T}{2\zeta} + \frac{\zeta}{2\Gamma T}$.

On this light field is superimposed a coherent field of constant intensity ρ_c so that the total generating function is $G_\rho(s; T) = G_t(s; T) \exp(-sS)$ where $S = \rho_c T$. Now with (III.1.15e), we obtain

$$p(n; T) = \sum_{m=0}^{n} (-1)^{(n-m)} \binom{n}{m} p_t(m; T) S^{n-m} \qquad \text{(III.2.3}f\text{)}$$

These distributions are usually characterized by the $\mathrm{E}[N^k] = \mu_k$ moments of N. In some situations, however, other coefficients are more significant. Such coefficients are briefly defined below and some relations with other moments are established.

Let $\phi_\rho(u)$ be the characteristic function expanded as follows

$$\phi_\rho(u) = \int_0^\infty \exp(iu\rho) \, W(\rho) \, d\rho = \sum_{k=0}^{\infty} \mathrm{E}[\rho^k] \frac{(iu)^k}{k!}$$

$$\overset{\text{def}}{=} \exp \left[\sum_{k=1}^{\infty} c_k \frac{(iu)^k}{k!} \right]$$

It is simply related to the probability generating function

$$G_N(z; T) = \sum_{n=0}^{\infty} z^n p(n; T) = \phi_\rho(i(1-z)) = \exp \left[\sum_{k=1}^{\infty} c_k \frac{(z-1)^k}{k!} \right]$$

$$= G_\rho(1 - z; T)$$

Generalized expressions of the characteristic functional and the probability generating functional for a process characterized by an integrated intensity (see Sections III.1.2– III.1.3a– for $g = 1$) $I(t; T) = \int_T g(t-u) \, dN(u)$ are studied in [42]. Generalized cumulants can also be derived.

From these equations, the integrated intensity moments $\mathrm{E}[\rho^k]$, the number moments $\mathrm{E}[n^k]$, the factorial moments f_k, the cumulants

c_k, and the factorial cumulants g_k are easily derived.

Now, because of the form of $\phi_\rho(u)$, the first-order derivative is sometimes more appropriate to work with

$$
\phi'_\rho(u) = G_N\left(\exp(iu); T\right) = \exp\left[\sum_{k=1}^{\infty} c_k \frac{(\exp(iu)-1)^k}{k!}\right]
$$

$$
= \exp\left[\sum_{k=1}^{\infty} c'_k \frac{(iu)^k}{k!}\right]
$$

where

$$
c'_k = \frac{d^k}{dv^k}\log\phi'_\rho(-iv)\Big|_{v=0} = \sum_{j=1}^{k} \frac{1}{j!}\frac{d^k}{dv^k}(\exp(v)-1)^j\Big|_{v=0} c_j
$$

$$
= \sum_{j=1}^{k}\sum_{m=1}^{j} \frac{(-1)^{j-m}}{m!(j-m)!} m^k c_j
$$

Let us illustrate this by an example in which a relation between g_k and f_k is shown. It is the thermal light with a Lorentzian spectrum [9,28,29], the kth factorial moment is given by

$$
f_k = E_n\left[n(n-1)\cdots(n-k+1)\right] = (-1)^k \frac{d^k}{ds^k}G_\rho(s;T)\Big|_{s=0} = E[\rho]^k
$$

Now, let $\Psi(s;T) = \log G_\rho(s;T)$. From (III.2.3b)

$$
\Psi(s;T) = -\sum_m \log(1 + sE[\rho]\lambda_m) = \sum_{k=1}^{\infty} (-1)^k \frac{s^k}{k!}E[\rho]^k \left(\sum_m \lambda_m^k\right)
$$

$$
\equiv \sum_{k=1}^{\infty} (-1)^k g_k \frac{s^k}{k!}
$$

so that $g_k = (k-1)! \, f_k \, \left(\sum_m \lambda_m^k\right)$ where λ_m are deduced from (III.2.3c,d). An interesting expression of g_k that is derived from a recurrence relation [44]

$$\sum_m \lambda_m^k = \int_{-\frac{T}{2}}^{-\frac{T}{2}} dt \; \gamma^{(k)} (t,t)$$

where

$$\gamma^{(k)} (t,t') = \int_{-\frac{T}{2}}^{-\frac{T}{2}} dx \exp\left(-\Gamma|t-x|\right) \gamma^{(k-1)} (x,t')$$

Generalization to a M–time joint distribution is now well known. Only some results of calculations for $M = 2$, which is the most useful situation for correlation measurements of thermal light source, are recalled here [18]. They can be derived from

$$G_\rho (s_1, s_2; T_1, T_2; \tau) = \frac{\exp\left(\Gamma (T_1 + T_2)\right)}{D(z_1, z_2; T_1, T_2)}$$

where the following notations are set for $i = 1,2$ [30,50,63]

$$D(z_1, z_2; T_1, T_2; \tau) = \left[\cosh\left(z_1 T_1\right) + \beta_1 \sinh\left(z_2 T_2\right)\right] \times$$
$$\left[\cosh\left(z_2 T_2\right) + \beta_2 \sinh\left(z_2 T_2\right)\right]$$
$$- \frac{\frac{E[\rho]}{\Gamma}^2 s_1 s_2 \sinh\left(z_1 T_1\right) \sinh\left(z_2 T_2\right) e^{-2\Gamma\tau - \Gamma(T_1 + T_2)}}{\sqrt{\left(1 + 2\frac{E[\rho]}{\Gamma} s_1\right)\left(1 + 2\frac{E[\rho]}{\Gamma} s_2\right)}}$$

where

$$z_i = \sqrt{\Gamma^2 + 2\Gamma E[\rho] s_i} \; ; \; \beta_i = \left[\left(1 + \frac{E[\rho]}{\Gamma} s_i\right)\left(1 + 2\frac{E[\rho]}{\Gamma} s_i\right)\right]^{-\frac{1}{2}}$$

For $\Gamma T_{1,2} \ll 1$

$$G_\rho(s_1, s_2; T_1, T_2; \tau) = \frac{1}{\mathcal{G}_\rho(s_1, s_2; T_1, T_2; \tau)}$$

$$\mathcal{G}_\rho(s_1, s_2; T_1, T_2; \tau) = 1 + \mathrm{E}[\rho](s_1 T_1 + + s_2 T_2)$$

$$+ \mathrm{E}[\rho]^2 T_1 T_2 s_1 s_2 (2 - \gamma_{\mathrm{I}}(\tau))$$

which yield [12]

$$\sum_{r_1=0,1} \sum_{r_2=0,1} (-1)^{r_1+r_2} A_{r_1 r_2} p(n_1 - r_1, T_1; n_2 - r_2, T_2; \tau) = 0$$

The matrix A being given by

$$A = \begin{pmatrix} A_{00} & A_{10} \\ A_{01} & A_{11} \end{pmatrix}$$

$$= \left(1 - \Gamma(\tau)^2\right) \begin{pmatrix} \frac{1+\langle n_1 \rangle + \langle n_2 \rangle}{1 - \Gamma(\tau)^2} + \langle n_1 \rangle \langle n_2 \rangle & \langle n_1 \rangle (1 + \langle n_2 \rangle) \\ \langle n_2 \rangle (1 + \langle n_1 \rangle) & \langle n_1 \rangle \langle n_2 \rangle \end{pmatrix}$$

with the boundary values:

$$p(k, T_1; 0, T_2, \tau) = \frac{A_{10}^k}{A_{00}^{k+1}}, \quad p(0, T_1; k, T_2; \tau) = \frac{A_{01}^k}{A_{00}^{k+1}}$$

The average photon numbers being $\langle n_i \rangle = \mathrm{E}[\rho] T_i$, i=1,2. A special distribution is obtained for $n_1 = 1$; $T_1 = T_2 = T$; $n_2 = n$,

$$p(1, T; n, T; \tau) = B_1^n + B_2 \sum_{m=1}^{n} B_1^{n-m} B_3^m \qquad \text{(III.2.4a)}$$

with

$$B_1 = \frac{A_{01}}{A_{00}}; B_2 = \frac{1}{A_{00}} \left(\frac{A_{01}}{A_{00}} - \frac{A_{11}}{A_{01}} \right); B_3 = \frac{A_{01}}{A_{00}} \qquad \text{(III.2.4b)}$$

and where $\langle n_1 \rangle = \langle n_2 \rangle = \langle n \rangle$.

The previous expression (III.2.4a) should not be compared with the $r(n, t = T; \tau)$, given by (III.1.17d). The two distributions are different because eq. (III.1.17d) is restricted to $T_1 \to 0$. This is the reason why we suggested in [28] to employ the latter method because of its simplicity when one deals with correlation measurements. For higher order time correlation measurements, a generalized time interval distribution can be calculated. It is denoted $w\left(t_1, ...t_M, t_i = 0, \tau_i | n = 1\right)$ where τ_i are the time intervals between the instants time of the point process $\tau_i = \vartheta_{i+1} - \vartheta_i$. In [47] the calculation of $w\left(t_1, t_2, t_i = 0, \tau_1, \tau_2 | n = 1\right)$ is given in detail. Here, we just recall a few results based on the utilisation of the derivatives of $G_\rho\left(s_1, s_2; T_1, T_2; \tau\right)$. For instance, from

$$w(t_1, t_2; \tau | n = 1) = \frac{1}{\mathrm{E}[\rho]^2} \frac{\partial^2}{\partial t_1^2} \frac{\partial^2}{\partial t_2^2} G_\rho\left(s_1, s_2; t_1, t_2; \tau\right)\bigg|_{s_1=s_2=1}$$

it is simple to prove

$$w(t_1, t_2; \tau | n = 1) = \frac{4\mathrm{E}[\rho]^2}{\left[1 + \mathrm{E}[\rho](t_1 + t_2) + \mathrm{E}[\rho]^2 t_1 t_2 \Lambda(\tau)\right]^3} \times$$
$$\left(\Lambda(\tau)^2 + \frac{6\left(1 - \Lambda(\tau)\right)\left(1 + \mathrm{E}[\rho]\Lambda(\tau)t_1\right)\left(1 + \mathrm{E}[\rho]\Lambda(\tau)t_2\right)}{\left[1 + \mathrm{E}[\rho](t_1 + t_2) + \mathrm{E}[\rho]^2 t_1 t_2\right]^2}\right)$$

where we set $\Lambda(\tau) = 2 - \gamma_\mathrm{I}(\tau) = 1 - \exp(-2\Gamma\tau)$.

For the special example $M = 2, E[\rho] = \Gamma, t_1 \to 0, \Gamma t \gg 1$, we obtain

$$w(0, t; t_i = 0, 0; \tau | n = 1) = \frac{2\mathrm{E}[\rho]}{(1 + \mathrm{E}[\rho]t)^3}\left[1 + \left(\gamma_\mathrm{I}(\tau) - 1\right)\frac{2 - \mathrm{E}[\rho]t}{1 + \mathrm{E}[\rho]t}\right]$$

Several probability distributions of interest for various parameters like the bandwidth, signal–to–noise ratio, ... are given in [47].

Now these results can be applied to detection purposes. As already noticed, the photon counting processing is the simplest (and often very efficient) method for the applications for communications and spectroscopy as well.

The calculations of detection performance are illustrated by plotting two typical figures. They are the ROC and $P_{er}(S)$ curves [14], in the case of the detection of coherent signal (average number of photons S) in a thermal background (average number of photons N_0) [48]. The thermal light is of an exponential second–order time correlation function of arbitrary correlation time τ_c.

These detection criteria have been defined and analyzed in Sect. II. The needed equations for computing Q_0, Q_d, P_{er} for binary testing, are given in subsection (II.1.1b) combined with eq. (III.2.3f).

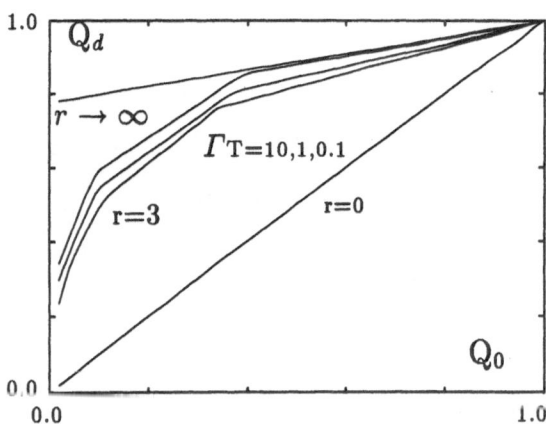

Fig. III.4. Receiver Operating Characteristic indexed with the r signal to noise ratio: processing with *relaxed* distribution. The average number of noise photons is $N_0 = 0.5$. For $r = 3$, $\Gamma T = 0.1, 1, 10$.

From Fig. III.4, it is observed that the performances improve when r increases. However, for $r \to \infty$, $Q_d(Q_0) \neq 1$ because of the time integration due to finite coherence time of the field. Moreover, we notice on the next figure that the performances improve when processing with $q(n)$ instead of $p(n)$ for several values of S.

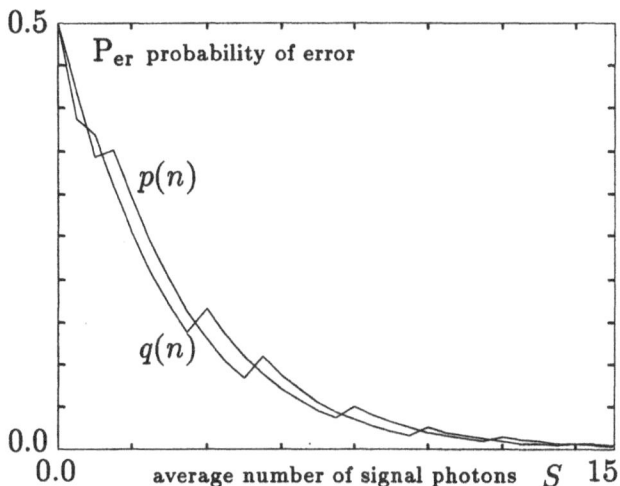

Fig. III.5. Minimum probability of error versus average signal photon number. The curve quoted $p(n)$ is obtained with *relaxed* distribution processing, $q(n)$ with *triggered* distribution processing. $\Gamma T = 2.0; N_0 = 0.5$

III.2.3 Time Intervals Processing

Instead of measuring the number of photons in a given time $[0, T]$ as done in the previous subsection, it might be useful in some cases to measure the time interval between photoelectrons within the same time duration $[0, T]$. To show how to calculate the detection performance, let us consider a very simple application : detection of a signal in noise, both being assumed coherent.

Let $t \in [0, T]$ be the r.v. of the nth occurence time of a Poisson process

(i) H_0: Poisson process of parameter λ_0; $p_0 \equiv v_0 (t|n)$

(ii) H_1: Poisson process of parameter $\lambda_1 = \lambda_s + \lambda_0$; $p_1 \equiv v_1(t|n)$

The LRT is equivalent to: $t \underset{H_1}{\overset{H_0}{\underset{<}{\gtrless}}} \mu_n$ [60].

Hence $Q_i = \int_0^{\mu_n} v_i(t|n)\, dt \quad (i = 0, 1) \cdot$

For $T \gg \mu$

$$Q_d \equiv Q_1 = 1 - (1 - Q_0)^{1+r} \tag{III.2.4}$$

where $r = \dfrac{\lambda_s}{\lambda_0}$. Here again and as seen in Figs. IV.3–4, Q_d increases when r increases.

III.3 Information Performance

In the following, we only consider *discrete memoryless* optical channels. The input is a random variable X taking $M + 1$ values $0, 1, ..., M$ with the *a priori* probability $\alpha_0, \alpha_1, ..., \alpha_M$, respectively. The output can be the number $N(T) = n_1, n_2, ..., \infty$ of received photons within the time interval $[0, T]$. With the same modulation format at the input we may choose $N(T) = 0, 1$. The definitions of different modulation formats have already been given. In the following subsections, some parameters are introduced to characterize the channels from the information point of view: channel capacity, probability of error and cut-off rate are calculated and analyzed for OOK and PPM modulations in several useful cases. However, some technical questions, important for the design of the system such as synchronization, overlapping,..., are not considered.
A review of sensitivity and capacity properties of several photon receivers is given in [59,62].

III.3.1 Channel Capacity

We start with the fundamental result of point process observations already demonstrated (see e.g. [46,52,55,68,74]).

The Poisson channel considered is a continuous–time additive noise channel in which the channel output $Y_t = X_t + N_t, t \in [0,T]$ is a Poisson–type point process with intensity $\eta_t = \chi_t + \rho_0$ where $\chi = \chi(\theta, Y)$ is an encoding of the message $\theta_t, t \in [0,T]$ and $\rho_0 \geq 0$ is the channel noise intensity. The encoder intensity is assumed to be a non–negative, causal, nonanticipative, noiseless, instantaneous feedback of the channel output Y_t. For a function $\chi(\theta)$ that does not depend on Y, the channel has no feedback.

The information capacity is defined as

$$C_t = \sup_{\theta} \sup_{\chi} \frac{1}{T} I_T[\theta, Y] \qquad (\text{III.3.1})$$

where I_T is the average mutual information between θ and Y.

For a channel of no constraint on the coding, the information capacity is infinite.

For a peak constraint $0 \leq \chi_t \leq Q$, it is shown

$$C_t = \frac{\rho_0}{e} \left(1 + \frac{Q}{\rho_0} \right)^{1 + \frac{\rho_0}{Q}} - \rho_0 \left(1 + \frac{\rho_0}{Q} \right) \log \left(1 + \frac{Q}{\rho_0} \right) \qquad (\text{III.3.2})$$

This result has been reconsidered by Davis [52] who included the average constraint $E[\chi_t] \leq Q'$, and more recently by Frey [75] who studied the problem of mean-square constraint $E[\chi_t^2] \leq Q^2$. It is interesting to notice that for these types of constraints, the encoding capacity is equal to the information capacity [68]. Different capacity formulas are therefore obtained, particularly the values of the capacity of the noiseless channel $\rho_0 = 0$. However they all conclude that, if there is no restriction on χ, the information channel capacity is infinite.

Now, let us focus on some special channels characterized by the type of signal modulations.

III.3.1a OOK–Modulation

(1) X–channel

$$I(\alpha) = (1 - \alpha)\mathcal{H}(p) + \alpha\mathcal{H}(q) - \mathcal{H}(\mathbb{P}_2) \qquad (\text{III.3.3}a)$$

For $q = 1 - p$, the channel is called *symmetrical*. The maximisation of $I(\alpha)$ is therefore obtained for $\alpha = \frac{1}{2}$ and the channel capacity reads

$$C = \log(2) + \mathcal{H}(q) \qquad (\text{III.3.3}b)$$

For the channel, as depicted below, which is a binary photoncounting with soft decision ($0 \leq k < \infty$), the calculations are straightforward.

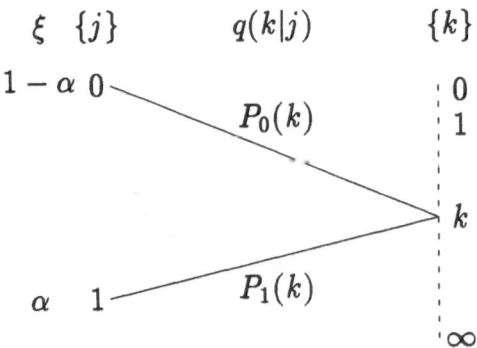

When the source input $X = j$, is given with an *a priori* probability ξ_j where $j = 1, ..., r$, the transition probability distribution for a soft transmission is $P_j(k)$ where $k = 0, 1, ..., \infty$. The mutual information between X and N is given by

$$I(\alpha, X; N) = (1 - \alpha)\mathcal{H}_0 + \alpha\mathcal{H}_1 - \mathcal{H}_{01}$$

where $\mathcal{H}_i = \sum_{n=0}^{\infty} P_i(n) \log P_i(n)$ with i=0,1 and

$$\begin{cases} \mathcal{H}_{01} = \sum_{n=0}^{\infty} \mathbb{P}(n) \log \mathbb{P}(n) \\ \mathbb{P}(n) = (1 - \alpha) P_0(n) + \alpha P_1(n) \end{cases}$$

The channel capacity will be $C_s^N = \max_{\alpha} I(\alpha, X; N)$. Now, suppose one restricts the received photon number $N(T) = n$ to be binary outputs such as $n = 0$ for $n \leq \mu$; $n = 1$ for $n > \mu$.
The mutual information can be written in the form

$$I(\alpha, \mu) = \mathcal{K}(\pi(\mu, \alpha)) - (1 - \alpha)\mathcal{K}(S_0(\mu)) - \alpha\mathcal{K}(T_1(\mu))$$

with the following notations

$$\begin{cases} \mathcal{K}(\pi(\mu, \alpha)) = -\pi(\mu, \alpha) \log(\pi(\mu, \alpha)) \\ \qquad\qquad - (1 - \pi(\mu, \alpha)) \log(1 - \pi(\mu, \alpha)) \\ \pi(\mu, \alpha) = (1 - \alpha) S_0(\mu) + \alpha T_1(\mu) \\ S_0(\mu) = \sum_{n=0}^{\mu} P_0(n) \\ T_1(\mu) = \sum_{n=0}^{\mu} P_1(n) \\ T_0(\mu) = 1 - S_0(\mu) \end{cases}$$

The capacity will be $C_h^N = \max_{\alpha,\mu} I(\alpha, \mu) = \max_{\mu} C^*(\mu)$. The maximum over α exists and is reached for

$$\alpha^*(\mu) = \frac{S_0(\mu)\left[1 + \exp(\chi(\mu))\right] - 1}{\left[1 + \exp(\chi(\mu))\right]\left[S_0(\mu) - T_1(\mu)\right]}$$

where

$$\chi(\mu) = \frac{\mathcal{K}(T_1(\mu)) - \mathcal{K}(S_0(\mu))}{T_1(\mu) - S_0(\mu)}$$

is a decreasing function from $\chi(0)$ to $-\infty$. A graphical study of $C^*(\mu)$ is convenient.

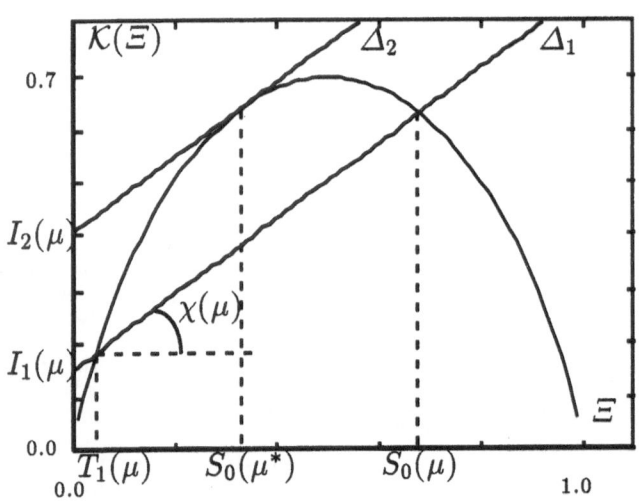

Fig. III.6. Graphical representation of $\mathcal{K}(\Xi)$ and $C^*(\mu) = I_2(\mu) - I_1(\mu)$.

With the help of Fig. III.6, it is shown that

$$C^*(\mu) = I_2(\mu) - I_1(\mu)$$
$$= \log\left[1 + \exp\left(\chi(\mu)\right)\right] - \chi(\mu) + \chi(\mu) S_0(\mu) - \mathcal{K}(S_0(\mu))$$

Hence the maximum is attained for

$$\chi(\mu^*) S_0(\mu^*) = \mathcal{K}(S_0(\mu^*))$$

with $\chi(\mu) \geq 0$.

This is approximately verified ($\forall \mu$, $T_1(\mu) \leq S_0(\mu)$) for $\mu^* \leq \mu_0$ for which $\mathcal{K}(T_1(\mu_0)) = \mathcal{K}(S_0(\mu_0))$.

(2) Z–channel

The *a priori* probability vector can be written as $Q = (\alpha, 1-\alpha)^t$.

The $P(Y|X) = \begin{pmatrix} 1 & 0 \\ 1-p & p \end{pmatrix}$ is the probability transition matrix and

the mutual information is given by $I_Z(\alpha; X, Y) = \alpha \mathcal{H}(p) - \mathcal{H}(\alpha p)$

where $\mathcal{H}(p) = p \log(p) + (1-p) \log(1-p)$ is the binary entropy

function. To calculate the capacity, we set the derivative of the

mutual information versus α equal to 0:

$$\frac{\partial}{\partial \alpha} I_Z(\alpha; X, Y) = 0 \longrightarrow \mathcal{H}(p) = \log\left(\frac{\alpha^* p}{1 - \alpha^* p}\right)^p$$

Therefore

$$C_Z(p) = I_Z(\alpha^*; X, Y) = \log\left(1 + p(1-p)^{\frac{1-p}{p}}\right)$$

Let us consider the Z–channel for which $p_0(n) = \delta(n)$. A closed

form for C_s^N can then easily be derived. It is shown that only one

maximum exists, obtained as before by setting the derivative with

respect to α equal to zero.

This occurs at $\alpha = \alpha^* = \dfrac{1}{1-p_1(0)+\exp\left(\frac{p_1(0)p_1(0)}{1-p_1(0)}\right)} \xrightarrow[p_1(0) \to 0]{} \dfrac{1}{e}$ which

leads to

$$C_s^N = C_Z\left(1 - p_1(0)\right) \qquad\qquad \text{(III.3.3c)}$$

(3) Comparison

It is interesting to compare from the capacity point of view

several types of channel that we already mentioned. We consider:

(0) Holevo bound

(i) photoncounting channel,

(ii) *quasi–classical* channel,

(iii) *quantum* channel,

(iv) *phase* channel.

The Holevo bound is derived from Eq.(II.3.2) with $\xi_i = \frac{1}{2}$. The eigenvalues of $|\psi_0\rangle\langle\psi_1| + |\psi_1\rangle\langle\psi_1|$ are calculated as in Sect. (II.2.1) so that

$$C^b = \log 2 - \frac{1}{2}\Big((1+\gamma)\log(1+\gamma) + (1-\gamma)\log(1-\gamma)\Big)$$

where $\gamma = \langle\psi_0|\psi_1\rangle = e^{-\frac{S}{2}}$ for pure real coherent states. For the last three models, the channel is a symmetrical one and characterized by a probability transition respectively given by

$$p_c = \int_{\sqrt{S}}^{\infty} \frac{1}{\sqrt{2\pi}} \exp(-\frac{x^2}{2})dx,$$
$$p_q = \frac{1}{2}(1 - \sqrt{1 - e^{-S}}),$$
$$p_\phi = \int_{-\frac{\pi}{2}}^{-\frac{\pi}{2}} p(\phi)d\phi$$

$$(III.3.3d, e, f)$$

$p(\phi)$ being given by eqs.(IV.13a,b).

For the case (i), the channel is of Z–type and the capacity was calculated in Eq. (III.3.3c).

For the other cases, the capacity in nats, is given by

$$C = \log(2) + p_i \log(p_i) + (1 - p_i) \log(1 - p_i), \qquad (i = c, q, \phi).$$
$$(III.3.3g)$$

The corresponding curves versus the average photon number $S = |\mu|^2$ are displayed in next figure. Notice that for (i)–(iv), the modulation is PSK.

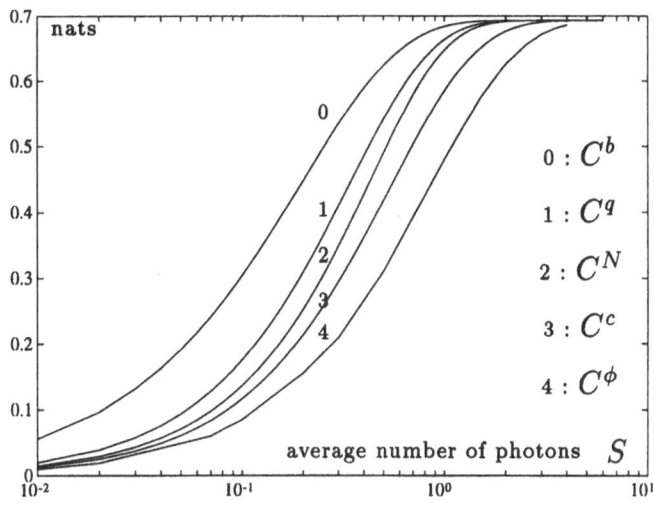

Fig. III.7. Capacity versus S for various channels.

It is clearly seen that the receiver based on optimum quantum detection is the closest to the upper bound. It is superior to the counter, the homodyne receiver and the phase receiver.

III.3.1b PPM–Modulation

(1) Soft Decision

Let $x \in A_X = (0,1)$ and $y \in A_Y = \mathbb{N}^M = \{n_i\}_{i=1}^M$. The channel capacity can be written, with $\zeta_i = \frac{1}{M}$, $\forall i$

$$C_s = \sum_{x \in A_X} \sum_{y \in A_Y} \frac{1}{M} \text{Prob}\,(y|x) \log \frac{\text{Prob}\,(y|x)}{\sum_{\xi \in A_X} \frac{1}{M}\text{Prob}\,(y|\xi)} \qquad \text{(III.3.4}a\text{)}$$

With the independence assumption

$$p\,(y|x_i) = \prod_{\substack{j=1 \\ j \neq i}} p_0\,(n_j)\,p_1\,(n_i)\,, \ \forall i$$

it is shown

$$C_s = \log M - \sum_{\{n_i\}_{i=1}^{M}=0}^{\infty} \left\{ p_1(n_1) \prod_{j=2}^{M} p_0(n_j) \log \left(\sum_{k=1}^{M} \frac{\Lambda(n_k)}{\Lambda(n_1)} \right) \right\}$$

$$(\text{III.3.4}b)$$

where $\Lambda(n_k)$ is the likelihood ratio (see III.1.1).

For a Z–channel

$$C_s = [1 - p_1(0)] \log M \qquad\qquad (\text{III.3.4}c)$$

(2) Hard Decision

The channel is defined by the following transition probabilities $\forall i = 1, ..., M$

(i) $i = j$, $p \overset{\text{def}}{=} \text{Prob}(Y = i | X = j)$

(ii) $i \neq j$, $q \overset{\text{def}}{=} \text{Prob}(Y = i | X = j)$

(iii) $\forall j, j \in \mathbb{N}^M$, $\epsilon \overset{\text{def}}{=} \text{Prob}(Y = \epsilon | X = j)$

(iv) $p + (M-1)q + \epsilon = 1$

$$p = \sum_{n=1}^{\infty} p_1(n) \left[\sum_{k=0}^{n-1} p_0(k) \right]^{M-1}$$

$$q = \sum_{n=1}^{\infty} p_0(n) \left\{ \left[\sum_{k=0}^{n-1} p_0(k) \right]^{M-2} \sum_{k=0}^{n-1} p_1(k) \right\}$$

Therefore, it is easily proved

$$C_h = (1 - \epsilon) \log M - (1 - \epsilon) \log(1 - \epsilon) + p \log p + (M-1) q \log q$$

It is demonstrated in [67] that $M \gg 1$, $C_h \leq p \log M$. Notice that for $q = 0$, $p = 1 - \epsilon = 1 - p_1(0)$, we obtain $C_h = C_s$.

(3) Two Results

It is important to notice the following results:

1. *Poisson distributions*:

For a weak noise channel, with an average power constraint, and a bandwidth constraint, it is proved that there exists M^*, the optimum alphabet size of M, such that

$$- \quad S \to 0, \quad M^*_{N_0=0} \sim e^{\frac{2}{S}}, \quad M^*_{N_0 \ll 1} \sim e^{\frac{2(1+2N_0)}{S}}$$

$$- \quad S \to \infty, \quad N_0 \ll 1, \quad M^* \sim 3 \quad [66].$$

2. *Coherent signal in a thermal noise of long coherence time*:

This example was extensively studied by Pierce et al [45,53]. It has been proved that the C_s soft decoding channel capacity is bounded by $C_s^m = \log\left(1 + \frac{1}{N_0}\right)$ for an average power constraint.

This result can be generalized to an arbitrary coherence time for an exponential second order time correlation function by using the Chernov bounds method [67]. Let us briefly explain how to obtain such generalization.

Denoting $G_i(z) = \sum_{k=0}^{\infty} p_i(k)z^k$, $(i = 0,1)$ the generating functions associated with the distributions $p_i(k)$ for a binary channel with $|z| \leq \xi_i$ and where $\xi_i \geq 1$ are the abscissa of convergence of $G_i(z)$. For a wide class of light fields, we have

$$\begin{cases} G_1(z) = G_0(z) \exp\left(-Sg\left(z\right)\right) \\ g(1) = 0 \\ \frac{dg(z)}{dz}\big|_{z=1} = -1 \end{cases}$$

where $p_0(k)$ is the distribution characterized by the average photon number N_0 and $p_1(k)$ is the distribution characterized by the average photon number $N_1 = N_0 + S$.

For example, $G_0(z) = \sum_{k=0}^{\infty}(1-\nu_t)\nu_t^k z^k = \frac{1-\nu_t}{1-\nu_t z}$ with $\nu_t = \frac{N_0}{1+N_0}$ for a thermal light and

$$G_1(z) = \sum_{k=0}^{\infty} p_0(k)e^{-S(1-\nu_t)} L_k\left(-\frac{S(1-\nu_t)^2}{\nu_t}\right) z^k$$

$$= G_0(z)\exp\left(-S(1-\nu_t)\frac{1-z}{1-\nu_t z}\right)$$

where $z \in [1, \frac{1}{\nu_t}[$ for a superposition of a coherent light and an independent thermal light. The Chernov's bound [7] reads here

$$\sum_{k=n}^{\infty} p_0(k) \leq \sum_{k=n}^{\infty} z^{n-k} p_0(k) \leq z^{-n} G_0(z)$$

Now, recall the formula established for the probability of error in the case of PPM–modulation (II.1.4)

$$P_{er} \leq \min_{\substack{\delta \in [0,1] \\ z \in [1,\xi_0]}} \sum_{l=0}^{\infty} p_1(l) \left[1 - \left(1 - \sum_{k=l}^{\infty} p_0(k)\right)^{M-1}\right]$$

$$\leq M^\delta \sum_{l=0}^{\infty} p_1(l) \left(\sum_{k=l}^{\infty} p_0(k)\right)^\delta$$

$$\leq G_0^\delta(z) G_0(z^{-\delta}) \, e^{\delta \log M - Sg(z^{-\delta})}$$

where we used

$$1 - (1-A)^{M-1} < M^\delta A^\delta, \; \forall \delta \in [0,1], \; \forall A \in [0,1]$$

Moreover, it is simple to prove as long as the rate

$$R_M \stackrel{\text{def}}{=} \frac{\log M}{S} \leq \log \xi_0$$

an *optimum* value M^* of M exists for which $P_{er} \rightarrow 0$. Therefore the maximum value of R_M is indeed the channel capacity. So, we may state

$$C = \log \xi_0 = \log \frac{1}{\nu_t} = \log \left(\frac{N_0}{1 + N_0} \right)$$

Let us apply this theory to the case of the superposition of the coherent light and the thermal light of exponential time autocorrelation function $\gamma(\tau) = \exp(-\Gamma|\tau|)$ with arbitrary coherence time $\frac{1}{\Gamma}$, the corresponding abscissa of convergence are shown to verify (see III.2.3d)

$$\begin{cases} x_i \tan(x_i) = 0 \\ \zeta_i = \frac{W}{W^2 + x_i^2} N_0 \\ \xi_0 = 1 + \frac{1}{\max\limits_i (\zeta_i)} \end{cases}$$

with $W = \frac{\Gamma T}{2}$, and where T is the time duration of measurement and N_0 the average value of photon number of the thermal component.

1. $\Gamma T \ll 1$, $x_i \sim W$ $\rightarrow \zeta_i \sim \frac{N_0}{1+W}$
2. $\Gamma T \gg 1$, $x_i \sim \frac{\pi}{2}$ $\rightarrow \zeta_i \sim \frac{N_0}{W}$

From the above discussion, we conclude that

$$C_s \leq \log \left(1 + \frac{1 + \frac{\Gamma T}{2}}{N_0} \right)$$

showing, for finite N_0

- $\Gamma T \rightarrow \infty$: Poisson channel
- $\Gamma T \rightarrow 0$: Pierce result

III.3.2 Cut-Off Rate

We briefly recall the coding theorem [6,7,40] for a discrete memoryless channel without cost constraint. For each pair $x_1, x_2 \in A_X$, we consider

$$\begin{cases} J(x_1, x_2) = \sum_{y \in A_Y} \sqrt{p(y|x_1)\, p(y|x_2)} \\ J_0 = \min_{Q(x)} \left\{ E[J(x_1, x_2)] \right\} \\ R_0 = \log(J_0) \end{cases}$$

For a binary channel where $A_X = (0, 1)$, $p(y|0) \equiv p_0(n)$, $p(y|1) \equiv p_1(n)$, the following expressions are obtained.

(1) Soft decision

As is shown hereafter, this cut–off is interesting because it is a limit for the hard decision processing. The calculation is straightforward

$$R_0^s = \log \frac{2}{1 + \sum_{n=0}^{\infty} \sqrt{p_0(n)\, p_1(n)}}$$

(2) Hard decision

The needed distributions are all already given in previous subsection. Here again, the calculations are very simple

$$R_0^h = \log \frac{2}{1 + \sqrt{S_0(\mu)\, T_1(\mu)} + \sqrt{S_1(\mu)\, T_0(\mu)}}$$

(3) Some general results

The hard decision performances are always worse than those obtained from the soft decision. To prove this, it is sufficient to apply the Schwarz inequality:

$$J_0^s = \sum_{n=0}^{\infty} \sqrt{P_0(n) P_1(n)} \leq \sqrt{\sum_{n=0}^{\mu} P_0(n) \sum_{n=0}^{\mu} P_1(n)}$$

$$+ \sqrt{\sum_{n=\mu+1}^{\infty} P_0(n) \sum_{n=\mu+1}^{\infty} P_1(n)}$$

$$= \sqrt{S_0(\mu) T_1(\mu)} + \sqrt{S_1(\mu) T_0(\mu)}$$

$$= J_0^h$$

Therefore, we can conclude $\forall \mu$, $R_0^s \geq R_0^h$

In the M–ary PPM signalling as proposed by Pierce for optical communication [45,53], the time axis is divided into M time slots and the light pulse is only present during the mth slot corresponding to the mth symbol. At the detector, synchronized to the PPM modulation , we receive $\{n\}_{i=1}^M$ photoncounts which are M independent random variables. If the message m_i is sent, let $p_0(n)$ be the probability that the light pulse is in the jth time slot for $j \neq i$. Questions concerning time synchronization are extensively studied in [38] (pp. 329–367).

Let $p_1(n)$ be the probability that the light pulse is in the jth time slot for $j = i$. The expression for the cut–off for a soft decoding receiver is given by [49,56,57,65]

$$R_0 = -\log \min_{\{Q_m\}} \sum_{\{n\}_{i=1}^M} \left(\sum_{m=1}^{M} \sqrt{p\left(\{n\}_{i=1}^M | m\right)} Q_m \right)^2$$

$$= \log M - \log \left(1 + (M-1) \left(\sum_{k=0}^{\infty} \sqrt{p_0(k) p_1(k)} \right)^2 \right)$$

For Poisson distributions, the cut–off rate is [56]

$$R_0 = \log M - \log \left[1 + (M-1) \exp \left(-\left(\sqrt{S+N_0} - \sqrt{N_0} \right)^2 \right) \right]$$

An analysis of this formula was made by Chan [51]. We recall the main results

— With bandwidth constraint

     – for $N_0 = 0$, $M \to \infty$, $R_0 \to \log 2$

     – for $N_0 \gg \log(M)$, $R_0 \sim S \sqrt{\dfrac{\log(M)}{N_0}}$

— With average power constraint and finite bandwidth, M^* is such that

     – $S \ll 1$, $M^* \sim \dfrac{2 \left(1 + \sqrt{2N_0} \right)}{S}$

     – $S \sim 1 + \sqrt{2N_0}$, $M^* \sim 3$.

These conclusions are only in qualitative agreement with those derived from the capacity analysis [66]. Finally, this subsection on photoncount processing is closed with an important theorem which is stated and its demonstration just outlined.

For the detection of coherent signal in thermal noise of long coherence time, the cut-off rate is bounded.

For $S \ll 1$, $R_0 \sim \frac{S}{4\nu_t(1+\nu_t)}$.

To find the bound for $S \gg 1$, we proceed as follows:

– first, notice that $R_0 \geq -\dfrac{1}{S} \min\limits_{z \in [1, \frac{1}{\nu_t}[} \left\{ G_0(z) G_1 \left(\dfrac{1}{z} \right) \right\}$

– then,

$$\left(\sum_{k=0}^{\infty} \sqrt{p_0(k)\, p_1(k)} \right)^2 \geq (1-\nu_t)^2\, e^{-S(1-\nu_t)} \times$$

$$\sum_{k=0}^{\infty} \nu_t^{2k} L_k \left(-\frac{S(1-\nu_t)^2}{\nu_t} \right)$$

– use now the inequality verified by the Laguerre polynomials for $x \geq 0, i \geq j$, that is $L_i(-x) \geq L_j(-x)$, to obtain

$$\left(\sum_{k=0}^{\infty} \sqrt{p_0(k)\, p_1(k)}\right)^2 \geq \exp\left(-\frac{S}{1+2N_0}\right)$$

– finally,

$$R_0 \leq \frac{1}{1+2N_0} \leq C$$

Comparison of the previous results show that the cut–off rate and the capacity points of view are quite different. The capacity criterion appears very difficult to attain.

Now, as already mentioned in III.2.2b, the time interval processor is often a preferable mode of operation because of its simplicity. Furthermore, in some situations, it performs better than conventional photon counting processing [72]. To illustrate this situation, let us consider two types of light of different second–order properties:

(i) thermal light exhibiting *bunching*,

(ii) squeezed light exhibiting *antibunching*

and examine the effects on the cut-off rate for these two specific examples.

Consider a Z–channel for which a correct detection of a "0" is always possible and the only error comes from the detection of the symbol "1". Assume that the receiver measures the time interval between photoelectrons. The transition probabilities are $p(0|0) = \delta(t)$; $p(1|1) = z(t)$, the time interval t being such that $t < S$. For soft–decoding, for which the fundamental limit is attained, the cut–off rate takes the form

$$R_0^s(S) = \log \frac{2}{1 + \int_0^\infty dt \sqrt{\delta(t) z (S - t)}}$$

The first case (i) (Fig. III.8), corresponds to the source of information is a thermal light of long coherence time, for which $z(t) \equiv v(t)$ given by (III.1.18a) yielding the curve denoted R_v, (resp. $w(t)$ given by (III.1.18b) yielding R_w). It is clearly seen that for $S \gg 1$, processing with life distribution yields higher performance. This is due to *bunching* effect which allow for $t \ll 1$ (i.e. $S \gg 1$), to have $w(S) \geq v(S)$.

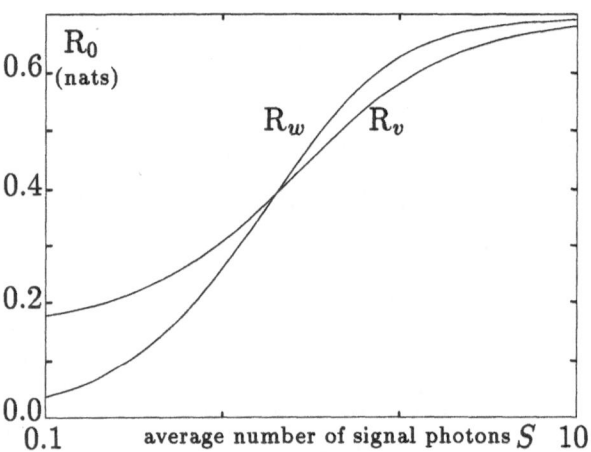

Fig. III.8. Thermal signal. Cut-off rate versus S the average number of the received photons. R_v: processing with the waiting time distribution. R_w: processing with the life time distribution.

Now, in the second case (Fig. III.9), the information is carried out by squeezed light for which $z(t)$ is $v(t)$ given by (III.1.20d), (resp. $w(t)$ given by (III.1.20e)). Because S depends on the degree of squeezing, this parameter must be optimized (e.g. by minimizing the probability of error).

Such a constraint on information criterion has already been considered for different purpose [10]. We observe that $\forall S$ processing with waiting time distribution leads to higher performance. This is due to the *antibunching* effect which tends to force the photons apart. More details of the calculations are given in [71].

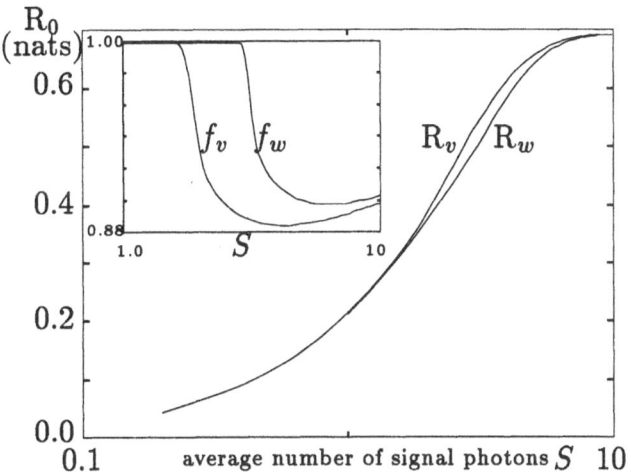

Fig. III.9. Coherent squeezed light. Cut-off rate versus S the average number of the received photons. R_v: processing with the waiting time distribution with minimization of the probability of error for $f_v(S)$. R_w: processing with the life time distribution with minimization of the probability of error for $f_w(S)$.

III.4 References

[1]   J.A. McFadden: Jour. Roy. Soc. **B 24** 364 (1962).
[2]   R. Glauber: Phys. Rev., **131** 2766 (1963).
[3]   P.L. Kelley, W.H. Kleiner: Phys. Rev. A, **30** 844 (1964).
[4]   L. Mandel, E. Wolf: Rev. Mod. Phys. **35**, 231 (1965).
[5]   A. Papoulis: *Probability, Random Variables and Stochastic processes* (Mc Graw-Hill, New York 1965).
[6]   J. Wozencraft, I. Jacobs: *Principles of Communication Engineering* (John Wiley & Sons, New York 1965).

[7]   R.G. Gallager: *Information Theory and Reliable Communication* (Academic Press, New York 1965).

[8]   I.S. Gradshteyn, I.M. Ryzhik: *Table of Integrals, Series and Products* (Academic Press, New York 1965).

[9]   R. Glauber: *Physics of Quantum Electronics*, ed. P. Kelly, B. Lax and O. Tannenwald (McGraw–Hill, New York 1966).

[10]   T.L. Gabriele: I.E.E.E. **IT–12** 484 (1966).

[11]   D.R. Cox, P.A.W. Lewis: *The statistical analysis of series of events* (Methuen, London 1966).

[12]   F.J. Beutler, O. Leneman: Acta Math. **116** 159 (1966).

[13]   G. Bédard: Phys. Rev. **161** 1304 (1967).

[14]   H.L. van Trees: *Detection, Estimation and Modulation Theory* (John Wiley & Sons, New York 1968). Part I.

[15]   J.R. Klauder, E.C.G. Sudarshan: *Fundamental of Quantum Optics* (Benjamin, New York 1968).

[16]   B.R. Mollow: Phys. Rev., **168** 1896 (1968).

[17]   F.T. Arecchi: *Photocount distributions and field statistics*, (International School of Physics " Enrico Fermi ", Varenna 1968).

[18]   D.B. Scarl: Phys. Rev., **175** 1661 (1968).

[19]   T.E. Duncan: Inf. Contr. **13** 62 (1968).

[20]   T. Kailath: I.E.E.E. **IT–15** 350 (1969).

[21]   B. Picinbono, M. Rousseau: Phys. Rev. A, **1** 630 (1970).

[22]   B. Picinbono, C. Bendjaballah, J. Pouget: Jour. Math. Phys. **11** 2166 (1970).

[23]   J. Perina: *Coherence of Light* (van Nostrand Reinhold, London 1971).

[24]   E.B. Davies: Comm. Math. Phys. **22** 5 (1971).

[25]   G. Lachs: Jour. Appl. Phys. **42** 602 (1971).

[26]   I. Rubin: I.E.E.E. **IT–18** 547 (1972).

[27]   O. Macchi: *Processus ponctuels et coincidences*, Thèse de l' Université d'Orsay (1972).

[28]   C. Bendjaballah: *Analyse de champs optiques par les méthodes de comptage et de coïncidences de photons*, Thèse de l' Université d'Orsay (1972).

[29]   C. Bendjaballah, F. Perrot: Jour. Appl. Phys. **44** 5130 (1973).

[30]   S.K. Srinivasan, S. Sukanam, E.C.G. Sudarshan: Jour. Phys. A **6** 1910 (1973).

[31]   R. Loudon: *The Quantum Theory of Light* (Clarendon Press, London 1973).

[32]   E.V. Hoversten, D.L. Snyder, R.O. Harger, K. Kurimoto: I.E.E.E. **COM–22** 17 (1974).

[33] R.O. Harger: I.E.E.E. **AES–11** 629 (1975).

[34] D.B. Snyder: *Random Point Processes* (Wiley, New York 1975).

[35] H.P. Yuen, J.H. Shapiro: I.E.E.E. **IT–21** 125 (1975).

[36] H.P. Yuen: Phys. Rev. A, **13** 2226 (1976).

[37] C.W. Helstrom: *Quantum Detection and Estimation Theory* (Academic Press, New York 1976).

[38] R.M. Gagliardi, S. Karp: *Optical Communications* (Wiley , New York 1976).

[39] R.O. Harger: *Optical Communication Theory* (Dowden, Hutchinson & Ross, Stroudsburg, (Pennsylvania). 1977).

[40] R.J. McEliece: *The Theory of Information and Coding* (Addison–Wesley Pub. Co., Reading (USA). 1977).

[41] R.S. Lipster, A.N. Shirayev: *Statistics of Point Processes II. Applications* (Springer Verlag, New York 1977).

[42] J. Rice: Adv. Appl. Prob. **9** 553 (1977).

[43] P. Brémaud, J. Jacod: Adv. Appl. Prob. **COM–9** 362 (1977).

[44] B.E.A. Saleh: *Photoelectron Statistics* (Springer–Verlag, Berlin 1978).

[45] J.R. Pierce: I.E.E.E. **COM–26** 1819 (1978).

[46] Y. Kabanov: Theory of Prob. & Appl. **IT–23** 490 (1978).

[47] C. Bendjaballah: Jour. Appl. Phys. **50** 62 (1978).

[48] C.W. Helstrom: I.E.E.E. **IT–25** 490 (1980).

[49] D.L Snyder, I.B. Rhodes: I.E.E.E. **IT–26** 327 (1980).

[50] R. Barakat, J. Blake : Physics Rep. **60** 225 (1980).

[51] V.W.S. Chan: I.E.E.E. **NTC–3** 57.4.1.–57.4.8. (1980).

[52] M. Davis: I.E.E.E. **IT–26** 710 (1980).

[53] J.R. Pierce, E.C. Posner, E.R. Rodemich: I.E.E.E. **IT–27** 61 (1981).

[54] M.D. Srinivas, E.B. Davies: Optica Acta **28** 981 (1981).

[55] P. Brémaud: *Point Processes and Queues* (Springer–Verlag, Berlin 1981).

[56] R.J. McEliece: I.E.E.E. **IT–27** 393 (1981).

[57] J.L. Massey: I.E.E.E. **COM–29** 1615 (1981).

[58] A.S. Holevo: *Probabilistic and Statistical Aspects of Quantum Theory* (North–Holland, Amsterdam 1982).

[59] H. Paul: Review of Mod. Phys. **54** 1061 (1982).

[60] C. Bendjaballah, K. Hassan: Jour. Opt. Soc. Amer. **73** 1840 (1983).

[61] B. Yurke: Phys. Rev. A, **32** 311 (1985).

[62] Y. Yamamoto, H.A. Haus: Rev. of Mod. Phys. **58** 1001 (1986).

[63] M. Singh : Optica Acta **33** 855 (1986).

[64] M. Cini, M. Serva : Jour. of Phys. A: Math. Gen. **19** 1163 (1986).

[65] M. Charbit, C. Bendjaballah: I.E.E.E. **COM–34** 600 (1986).

[66] M. Charbit, C. Bendjaballah: Optics & Quant. Elect. **18** 49 (1986).

[67] M. Charbit, C. Bendjaballah: I.E.E.E. **COM–35** 122 (1987).

[68] A.D. Wyner: I.E.E.E. **IT–34** 1449 (1988).

[69] R. Vyas, S. Sing: Phys. Rev. A, **38** 2423 (1988).

[70] D.J. Daley, D. Vere–Jones: *An Introduction to the Theory of Point Processes* (Springer–Verlag, Berlin 1988).

[71] C. Bendjaballah, H. le Pas de Sécheval: Optics Comm. **74** 403 (1990).

[72] H. le Pas de Sécheval, C. Bendjaballah: *Coherence and Quantum Optics VI*, ed. by L. Mandel and E. Wolf (Plenum Press, New York 1990). p. 675.

[73] A. Barchielli: *Quantum Aspects of Optical Communications*, ed. by C. Bendjaballah, O. Hirota and S. Reynaud (Springer Verlag, Berlin 1991).

[74] A.O. Hero: *Proc. of the Conf. on Information Sciences and Systems* (John Hopkins, Baltimore, USA 1991).

[75] M.R. Frey: I.E.E.E. **IT–37** 244 (1991).

IV. Phase Operator

The problem of a phase operator, which obeys the laws of quantum mechanics, has inspired many physicists since Dirac first assumed that a Hermitian phase operator, canonically conjugate to the number operator, existed. On the other hand, the study of squeezed states of light has renewed interest in quantum phase operator measurement. Numerous studies have been devoted to this subject but several questions remain unsolved. It seems that no Hermitian operator corresponds to the phase of a quantum electromagnetic field.

One of the problems one can examine, is to find which phase probability distribution is relevant to the observation of quantum phase. We may also wonder what the quantum phase operator, analyzed in [1], stands for, how it is specifically related to the quantum phase observable introduced in [2] and extensively discussed in [5]. Another question relevant to the wave packet reduction postulate, is to describe the state of the quantum oscillator after having measured its phase.

All these questions are obviously essential from a theoretical point of view. However, they deserve to be studied from a practical point of view too. Indeed, in order to apply techniques such as heterodyning, to use squeezed light or phase shift keying modulation in optical experiments involving phase measurement, it is important to answer these questions and to determine the quantum limit performance for optical communications based for example on phase measurement.

Numerous authors have already proposed solutions which, however, differ significantly in many points, e.g. phase probability distribution and the commutator of photon number–phase operators. The problem comes essentially from the difficulty of defining an appropriate quantum phase operator dynamically conjugated to the photon number due to the fact that the shift $n \mapsto n+1$ of $n = 0, 1, 2, \dots$ cannot be inverted and hence does not generate a unitary group in the corresponding Hilbert space.

IV.1 Some Models

The first method proposed by Susskind and Glogower (see [1]) defines $\widehat{U} = \widehat{C} + i\widehat{S}$ as the phase operator via the sine and cosine operators $\widehat{S}, \widehat{C}$ such that the (non–normalizable) phase states are given by

$$|\Psi(\phi)\rangle = |\exp(i\phi)\rangle = \sum_{n=0}^{\infty} e^{in\phi}|n\rangle \qquad (\text{IV.1})$$

defined in terms of the number states $|n\rangle$.

This model provides a satisfactory description in the region of high energy states and leads to the p.o.m.

$$\widehat{\Pi}_\Delta(\phi) = \frac{1}{2\pi} \int_\Delta |\exp(i\phi)\rangle\langle\exp(i\phi)|d\phi \qquad (\text{IV.2})$$

with $\Delta = [\phi, \phi + 2\pi)$.

However, because the cosine and sine operators do not commute $[\widehat{C}, \widehat{S}] = -\frac{1}{2}|0\rangle\langle 0| \neq 0$, it is not possible to find a complete set of eigenvectors of $\widehat{C}$ and $\widehat{S}$ simultaneously. Therefore this description is only *approximate*.

In favor of this model, a remark made by [3] that

$$\hat{E} = \sum_{n=0}^{\infty} |n\rangle\langle n+1| \qquad (IV.3)$$

with eigenstate $|\phi\rangle = \frac{1}{2\pi}\sum_{n=0}^{\infty} e^{in\phi}|n\rangle$ is the appropriate operator although it is not hermitian ($\hat{E} \neq \hat{E}^{\dagger}$), if $\hat{E}$ could be adopted for the same reason that an *observable* $\hat{a}$ is accepted despite that $\hat{a} \neq \hat{a}^{\dagger}$.

Another approach is based on the maximum likelihood ratio criterion of the quantum estimation theory [6,10], to obtain the p.o.m. and the conditional probability $p(\phi|\theta)$ of measuring the phase estimate ϕ given that the true value is θ

$$\begin{cases} d\hat{\Pi}(\phi) = \frac{1}{2\pi}\exp(-i\hat{N}\phi)|\gamma\rangle\langle\gamma|\exp(i\hat{N}\phi)d\phi \\ p(\phi|\theta)d\phi = \langle\psi(\theta)|d\hat{\Pi}(\phi)|\psi(\theta)\rangle \end{cases} \qquad (IV.4)$$

where $\hat{N}|k\rangle = k|k\rangle$ and $|\gamma\rangle$ is defined in terms of projections of states of the system in $|\psi(\theta)\rangle = \sum_{n=0}^{\infty} \psi_n \exp(in\theta)|n\rangle$ a state function of the unknown parameter θ ($\rho(\theta) = |\psi(\theta)\rangle\langle\psi(\theta)|$), into $|k\rangle$. To demonstrate (IV.4), some elements of the quantum estimation theory are recalled in Sect. II.2.1. (see eqs. (II.2.3a–e)).

Obviously, this description strongly depends upon a special criterion of quantum estimation theory which implies that the p.o.m. is not optimal in either strategy, for instance Shannon's mutual information [19]. Nevertheless, to prove that this model is useful, we study in which conditions the operators $\hat{N}$ and $\hat{\Phi}$ are conjugate according to Kraus [4]. Two observables $\hat{A}$, and $\hat{B}$ of eigenvectors $\{|a_i\rangle\}$ and $\{|b_j\rangle\}$ in a Hilbert space of dimension d are called *complementary* if $\forall i, j |\langle a_i|b_j\rangle| = \frac{1}{\sqrt{d}}$.

Furthermore, the relations between optimal detection operator and phase shift measurement are analyzed [7].

Let $|\mu_k\rangle \equiv |\mu u^k\rangle = |\mu \exp\left(-i\frac{2\pi k}{d}\right)\rangle$ be phase modulated (mod. d) coherent states given with probability *a priori* $\left\{\zeta_k\right\}_{k=0}^{d-1}$ where $u = \exp\left(-i\frac{2\pi}{d}\right)$ and denote $\widehat{\Psi}_k = \exp\left(-ik\widehat{N}\frac{2\pi}{d}\right)$, the shift operator $|\mu_k\rangle = \Psi_k|\mu\rangle$. The Gram matrix $\underline{G}$ of elements $g_{ij} = \langle\mu_i|\mu_j\rangle$ is hermitian and circulant $g_{ij} = \exp\left(-\mu^2 + \mu_i^*\mu_j\right) = \exp[-\mu^2\left(1 - u^{j-i}\right)]$. From $\underline{G}|\beta_p\rangle = \lambda_p|\beta_p\rangle$, we obtain

$$|\beta_p\rangle = \frac{1}{\sqrt{d}}\left[1, u^{-p}, \cdots, u^{-pk}, \cdots, u^{-p(d-1)}\right]^\dagger \qquad (\text{IV.5})$$

for $0 \leq p \leq d - 1$ and $0 \leq k \leq d - 1$.

The eigenvalues are given by $\lambda_p = \sum_{k=0}^{d-1} g_{0k} u^{-pk}$ where $g_{0k} = \exp[-\mu^2\left(1 - u^k\right)]$.

We now define $\underline{X}$ so that $\underline{X}^\dagger \underline{X} = \underline{G}$, and using the spectral decomposition theorem, we set

$$\underline{G} = \sum_{p=0}^{d-1} \lambda_p|\beta_p\rangle\langle\beta_p|$$

$$\underline{X} = \sum_{p=0}^{d-1} \xi_p|\beta_p\rangle\langle\beta_p| \equiv \sum_{p=0}^{d-1} \sqrt{\lambda_p}\exp(i\theta_p)|\beta_p\rangle\langle\beta_p|$$

Let us examine in which case the p.o.m. $\widehat{\Omega}_m = |\omega_m\rangle\langle\omega_m|$, $\sum_m \widehat{\Omega}_m = \mathbb{1}$, are the optimum detection operators (o. d. o.). Given the results of the quantum detection theory (see Sect. II), detection probability Q_d is maximum for

$$
\begin{cases}
\underline{X} = \underline{G}^{\frac{1}{2}} \quad \text{because } x_{rs}x_{ss}^* = x_{rr}x_{sr}^* \\
\zeta_k = \frac{1}{d}, \quad \theta_p = 0 \\
Q_d = |x_{00}|^2 = \dfrac{1}{d^2} \left| \sum_{p=0}^{d-1} \sqrt{\lambda_p} \right|^2
\end{cases} \qquad \text{(IV.6}a, b, c)
$$

where we set $x_{rs} = \langle \beta_r | \underline{X} | \beta_s \rangle$. To evaluate the λ_p, we start with

$$
|\mu_k\rangle = \exp\left(-ik\widehat{N}\frac{2\pi}{d}\right)|\mu\rangle = \sum_{r=0}^{\infty} \exp\left(-ikr\frac{2\pi}{d}\right)|r\rangle\langle r|\mu\rangle \quad \text{(IV.7)}
$$

and taking into account that

$$
g_{0k} = \langle \mu | \mu_k \rangle = \sum_{r=0}^{\infty} |\langle r|\mu\rangle|^2 u^{kr} = \exp(-S) \sum_{r=0}^{\infty} \frac{S^r}{r!} u^{kr}
$$

where $\sum_{r=0}^{d-1} u^{kr} = 0$ if $k \neq 0$, d if $k = 0$, we arrive at

$$
\lambda_p = \sum_{k=0}^{d-1}\sum_{r=0}^{\infty} |\langle r|\mu\rangle|^2 u^{kr} u^{-pk} = d\exp(-S)\sum_{n=0}^{\infty} \frac{S^{p+nd}}{(p+nd)!} \quad \text{(IV.8)}
$$

Then as $x_{rs} = \dfrac{1}{d}\sum_{p=0}^{d-1} \sqrt{\lambda_p} u^{-ps} u^{pr}$ we readily obtain

$$
\langle n|\omega_m\rangle = \sum_{\ell=0}^{d-1} \xi_{\ell m} \langle n|\mu_\ell\rangle
$$

$$
= \frac{1}{d}\exp\left(-\frac{S}{2}\right)\sqrt{\frac{S^n}{n!}} \sum_{\ell=0}^{d-1}\sum_{k=0}^{d-1} \frac{1}{\sqrt{\lambda_k}} \exp\left[\frac{2i\pi}{d}\left(k(\ell-m)-\ell n\right)\right]
$$

$$
\text{(IV.9)}
$$

Therefore

$$
|\langle n|\omega_m\rangle| = \exp\left(-\frac{S}{2}\right)\sqrt{\frac{S^n}{n!}}\frac{1}{\sqrt{\lambda_n}} = \frac{1}{\sqrt{d}}\left[1 - \frac{S^d}{2(n+d)} + \cdots\right]
$$

$$
\text{(IV.10)}
$$

This result demonstrates that the observables $\hat{N}$ and $\widehat{\Omega_m}$ are complementary for $S^d \ll 4d$, that is verified for $S \leq 1$, $\forall d \geq 1$.

Let us analyze the asymptotical behavior with respect to d. As
$$\sum_{n=0}^{\infty} \frac{S^{p+nd}}{(p+nd)!} \sim \frac{S^p}{p!}, \text{ it can be seen that } \lim_{d \to \infty} \frac{\lambda_p}{d} = \exp(-\mu^2)\frac{\mu^{2p}}{p!}$$
which is a Poisson distribution.

From eq.(IV.8), we get $x_{r0} \equiv \langle \mu | \beta_r \rangle$. If we set $|\phi\rangle = \frac{1}{\sqrt{\frac{2\pi}{d}}}|\beta_r\rangle$, x_{r0}
can be rewritten as (with $\varphi = \frac{2\pi}{d}r$)

$$w(\varphi) = \langle \mu | \phi \rangle = \lim_{d \to \infty} \frac{1}{\sqrt{2\pi}} \sum_{p=0}^{d-1} \exp\left(-\frac{\mu^2}{2}\right) \frac{\mu^p}{\sqrt{p!}} \exp\left(-i\varphi p\right)$$

$$(IV.11)$$

where

$$|\phi\rangle = \frac{1}{\sqrt{2\pi}} \sum_{n=0}^{\infty} \exp\left(-in\varphi\right)|n\rangle \qquad (IV.12)$$

which is similar to eq.(IV.1). The probability distribution of the phase shift

$$p(\varphi) = |w(\varphi)|^2 = \frac{1}{2\pi}\left|\sum_{n=0}^{\infty} \exp\left(-in\varphi\right) \langle \mu | n \rangle\right|^2 \qquad (IV.13a)$$

takes the form

$$p_{cs}(\varphi) = \exp(-|\mu|^2) \sum_{p,q=0}^{\infty} \frac{1}{p!q!} \mu^{p+q} \cos\left[(p-q)\varphi\right] \qquad (IV.13b)$$

using (IV.10) in the case of a coherent state (CS).

Remark that, owing to the central limit theorem (see § III.1.3d), $|\langle \mu | n \rangle|^2$ can be approximated by a Gaussian form for $\mu \to \infty$, the distribution (IV.13b) will therefore exhibit a Gaussian behavior. The calculation of the phase distributions for other models is continued for thermal light (TS) for which

$$p_{TS}(\varphi) = \frac{1}{2\pi} \frac{1 - \nu^2}{1 - 2\nu \cos\varphi + \nu^2}, \qquad (-\pi \le \varphi \le \pi) \qquad \text{(IV.13c)}$$

where $\langle \mu | n \rangle = \sqrt{1 - \nu}\nu^{\frac{n}{2}}, n = 0, 1, ..., \infty$ with $\nu = \frac{|\mu|^2}{1+|\mu|^2}$. For a space of finite dimension d, we obtain [18]

$$p_{TS}(\varphi) = \frac{1}{2\pi} \frac{1 - \nu^2}{1 - 2\nu \cos\varphi + \nu^2} \frac{1 - 2\nu^{d+1}\cos((d+1)\varphi) + \nu^{2d+2}}{1 - \nu^{2d+2}}$$

$$\text{(IV.13c')}$$

For pure squeezed light (SS), the phase distribution is derived from (III.1.20b) with $S_c = |\mu|^2, c^2 = \frac{e^{2r}S_c}{\sinh(2r)}$

$$\langle \mu | n \rangle = \frac{1}{\sqrt{\cosh r}} \ e^{-\frac{S_c}{2}\left(1 + \tanh r\right)} \frac{\left(\frac{\tanh r}{2}\right)^{\frac{n}{2}}}{\sqrt{n!}} \ H_n(c) \qquad \text{(IV.13d)}$$

which is put into (IV.13a) for numerical computation.

These three distributions, which are symmetrical in φ, will be displayed in Fig. IV.1a together with the square root of the product of variances of n and φ in Fig. IV.1b. To compute the square root of this product in an approximately closed form, we may consider, as in the case of x and p variables (I.2.26), a joint distribution of the conjugated variables. The number of photons n and phase φ can be obtained from a Wigner distribution

$$W(n, \varphi) = \int du \ e^{\frac{i}{\hbar}nu} w(\varphi - \frac{u}{2}) w^*(\varphi + \frac{u}{2}) \qquad \text{(IV.14)}$$

which has been evaluated in the case of a highly squeezed state [8].

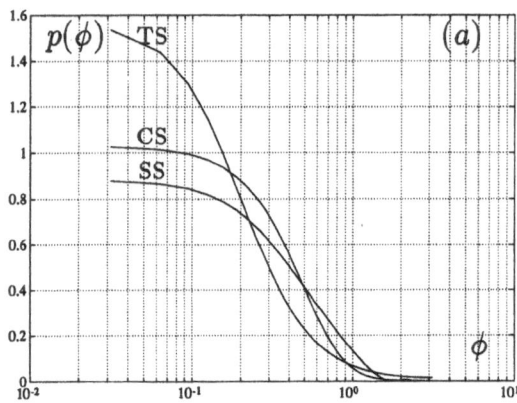

Fig. IV.1a. The curves display the phase probability distributions for coherent squeezed state (SS) with $f = 0.8555$, coherent state (CS) and thermal state (TS) versus the phase ϕ. The received number of photons is $S = 2$.

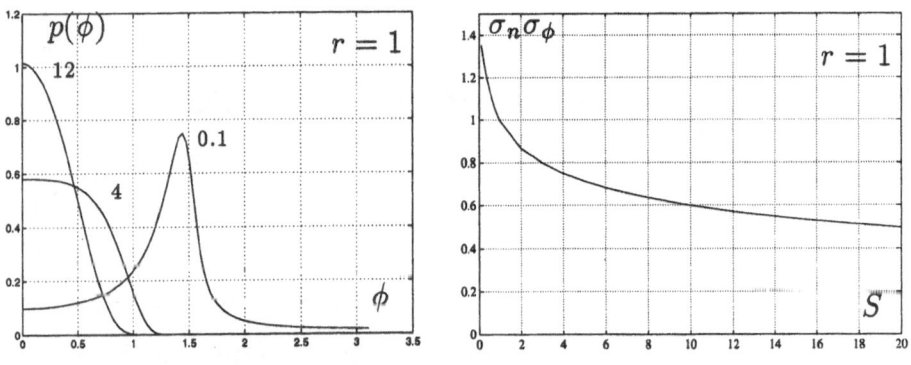

Fig. IV.1b. The product of the square root of the variances $\sigma_n \sigma_\phi$ are plotted versus the received number of photons S, σ_n being given by III.1.20c for the SS. In this case the curves $p(\phi)$, parametrized with $S = 0.1, 4, 12$, undergoe a bifurcation (see [8]).

Computation of $\sigma_n \sigma_\phi$ derived from eq. (IV.13b), for curve (CS), shows that $\sigma_n \sigma_\phi \to \frac{1}{\sqrt{2}}$. To compare with curve (SS), we set $r = \text{atanh}\left(\sqrt{\frac{(1-f)S}{1+(1-f)S}}\right)$ and $f \sim 1 - \frac{\sqrt{S}}{3}e^{-\sqrt{\frac{2S}{3}}}$. In this case, approximate calculations show that for $S \gg 1, \sigma_n \sigma_\phi \sim 0.6$ [17]. This asymptotic value is close to the value that one can obtain by extrapolating the values derived from the $p(\phi)$ (see curve SS of Fig. IV.1a). The deviation might be due to the method used to fix f and to some lack of precision in numerical computation. Now, if S are r are independent, as it is in Fig IV.1b, then $\sigma_n \sigma_\phi \to 0.5$.

The corresponding curve for the (TS) is continuously increasing (in the range of the S considered here), due to the very slow decrease of the thermal photoncounting distribution.

All the results obtained here seem in a satisfactory agreement with the models (e.g. [9,13,17]), except in the region of a weak S.

However, it must be pointed out that the probability distributions are not accurately normalized to 1, because of a lack of precision in the numerical computation.

In the third approach of the phase operator, recently proposed in [5], Pegg and Barnett introduce the phase operator from a finite–dimensional space

$$
\begin{cases}
\hat{\Phi}_d = \sum_{k=0}^{d} \theta_k^{(d)} |\theta_k^{(d)}\rangle \langle \theta_k^{(d)}| \\
|\theta_k^{(d)}\rangle = \frac{1}{\sqrt{d+1}} \sum_{n=0}^{d} \exp(in\theta_k^{(d)})|n\rangle \\
\theta_k^{(d)} = \theta_0 + 2\pi \frac{k}{d+1}
\end{cases}
\qquad (IV.15)
$$

The state–space dimension d may tend to infinity only after physical results such as expectation values have been calculated. Thus, defining a mapping between any function of operators $\widehat{\varPhi}_c$ and states $|\psi\rangle_c$ of the conventional Fock space $\mathcal{H}, \{|n\rangle\}$ and those defined in the truncated space $\mathcal{H}_d$ [12]

$$
{}_c\langle\theta_k|f\left(\widehat{\varPhi}_c\right)|\theta_k\rangle_c = \lim_{d\to\infty} \frac{\langle\theta_k^{(d)}|f\left(\widehat{\varPhi}_d\right)|\theta_k^{(d)}\rangle}{\langle\theta_k^{(d)}|\theta_k^{(d)}\rangle} \tag{IV.16}
$$

This mathematical construction is interesting because these phase states form a complete orthonormal basis.

Furthermore they yield, among others, the following properties

$$
\begin{cases}
\exp\left(i\widehat{\varPhi}_d\right)|0\rangle = e^{(d+1)\theta_0}|d\rangle \\[2mm]
\langle n|\cos^{2q}|n\rangle = \langle n|\sin^{2q}|n\rangle = \displaystyle\prod_{m=1}^{q} \frac{2m-1}{2m}
\end{cases} \tag{IV.17}
$$

which are very important and not given by the previous models.

As a simple application, let us briefly point out some remarks concerning the entropy of the probability distribution of coherent phase.

Instead of utilizing Shannon's entropy $S_\phi(n) = \log d$ which diverges for $d \to \infty$, due to the complementarity property (IV.10), it is suggested to work with the relative entropy $S_\phi(n) + \log q(d)$ which is bounded for $q(d) = \frac{1}{d}$. The choice of this *a priori* probability is motivated by the fact that the knowledge of the dimension of the space must be taken into account.

On the other hand, it is of interest to evaluate how thermalization of the state changes the entropy. When measuring the photon number, it is simple to prove that for $S \gg 1$ and $d \gg 1$,

$$S^N_{TS} = 1 + \log S \qquad \text{(IV.18)}$$

It is then derived that there is an entropy gain by thermalization of a coherent state $\Delta S^N = S^N_{TS} - S^N_{CS} \sim \log \sqrt{S}$, as $S^N_{CS} \cong \frac{1}{2}(1 + \log 2\pi S)$ for the CS. The Poisson distribution being approximated by a Gaussian distribution of mean S and variance S.

In order to do the same calculations for ΔS^ϕ, we use (IV.13c) to obtain , for $\nu \to 1$,

$$S^\phi_{TS} \geqslant \log 2\pi(1 - \nu^2) \cong 2\log 2 + \log \pi - \log S \qquad \text{(IV.19)}$$

so that after the phase normalization, one obtains

$$S \gg 1, \quad S^\phi_{CS} \cong \frac{1}{2}\left(1 + \log \frac{\pi}{2S}\right), \quad \Delta S^\phi = S^\phi_{TS} - S^\phi_{CS} < 0$$

The used approximate distribution fits very closely the dominant central curve of (IV.13c') but does not reproduces the side oscillations.

Remark that the entropy for phase measurement is a decreasing function for both variables $S > 1$, $d > 1$.

Finally, notice that the sum $\Delta_{TS} = S^\phi_{TS} + S^N_{TS} \cong 1 + \log \pi + 2\log 2 > \Delta_{CS}$, as required by the entropic uncertainty relation (I.2.3).

Another method that yield similar conclusions is based on the expansion of (I.2.10a) with respect to the thermal parameter θ (square root of the average photon number), using results of the thermofield analysis. However, these results are valid for a small θ only. This method has recently been used to study the case of a weakly squeezed state.

It is concluded that some information could be gained for high values of the squeeze parameter [16].

In fact, the point of view of truncating the Fock space, does not differ from the Susskind–Glogower model for the statistical properties at least as long as the states are normalizable.

However, other questions should be clarified [11,12], such as the restriction to *physical states* (physical states are those where all number moments are finite $\langle n^q \rangle < \infty$ $\forall q$), the nullity of the trace of all commutators, the fact that $\exp(i\widehat{\Phi}_d)|0\rangle = \exp((d+1)\theta_0)|d\rangle$ where d is supposed to tend to ∞ (a state of infinite energy), and the fact that normalization of the phase states for finite d and $d \to \infty$ take place in different spaces.

In order to avoid the difficulties due to the fact that the phase state is defined only on a subset of the entire Hilbert space and in order to have a superposition of all states ($n \to \infty$) as for the definition of coherent states, a recent modification has been proposed [20] in the form $|\phi\rangle = A(r) \sum_{n=0}^{\infty} \frac{\exp(-\frac{n}{\mu})}{n+r} |n\rangle$ which is normalized so that $A(r) \sim r$ and assuming both $r \ll 1$ and $\mu \gg 1$.

From another point of view, a different formulation which eliminates the asymptotic procedure by using the QND method [18] has also been proposed.

To measure the phase–operator, defined by an analytical representation in the unit disc [14], it has been suggested that the system of interest should be coupled with an apparatus and that the measurements should be made on a more acceptable physical observable of the apparatus, e.g. the angular momentum system in a finite dimensional Hilbert space.

The Hamiltonian for the system commutes with the angular operator of the apparatus and it is therefore a QND operation. The observable to be measured is the meter observable which is diagonal in the basis where the momentum is a pure differential operator. An initial state for the apparatus has also been proposed and the corresponding reduction transformation is found.

Finally, it has been concluded that the operator representing the phase of a quantum field cannot be separated from the measurement process. Thus, for a measurement scheme, a corresponding quantum phase operator can be defined (see also [15]).

To briefly conclude this section, let us consider the performances of some phase receivers in information transmission, the performance being characterized by the cut off rate criterion.

As already noticed, the cut off rate criterion is not as meaningful as the channel capacity criterion. This is partly because, for noiseless M–ary channel, different receivers, namely o.d.o. and number operators, lead to equal cut off. However, it is useful when

(i) comparing the performance at a first encoding approach,

(ii) comparing the performance for the same receiver but for different types of input signals.

In the following figure, it is clearly observed that the phase receiver is the worst compared to the o.d.o. and the homodyne. It is interesting to notice that the slopes of the variations versus the average signal photon number, are equal for weak S.

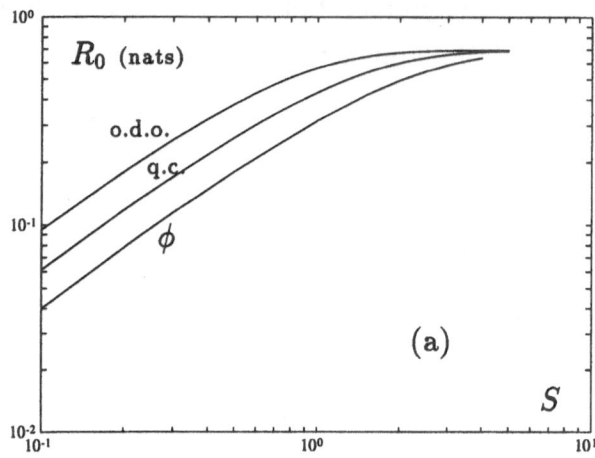

Fig. IV.2a. Cut off rate performance given by (II.3.3b) for phase shift keying channel with the (CS) model of phase with different receivers: the curve o.d.o. is plotted by using eqs. (II.3.18), the curve q.c. is obtained from (III.3.3d), the curve quoted ϕ from (III.3.3f)–(IV.13b), versus S the average signal photon number.

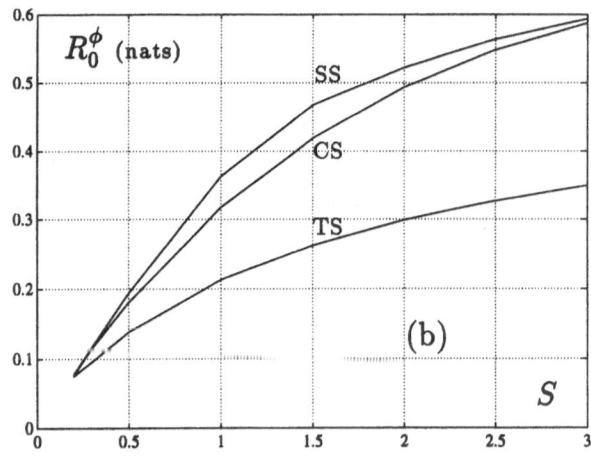

Fig. IV.2b. Cut off performance for the same channel in the case of three different phase models: (CS) coherent state, (TS) thermal state and (SS) squeezed state with $f = f_m(S)$ (see p. 136), versus S the average signal photon number.

From the above figure, it is noted that constructing a receiver with a squeezed phase state is interesting for $S \leq 3$. However, in that range of S, the resulting performance are not greatly improved compared with the coherent phase state, contrary to the case of the photoncounting processor. Calculation of performance for higher values of S and analysis of the results will be done in a future work.

IV.2 References

[1]   P. Carruthers, M.M. Nieto: Review of Modern Physics, **40** 411 (1968).

[2]   R. Loudon: *The Quantum Theory of Light* (Clarendon Press, London 1973)

[3]   J.M. Lévy-Leblond: Annals of Physics, **101** 319 (1976).

[4]   K. Kraus: Physical Review D, **37** 3070 (1987).

[5]   D.T. Pegg, S. M. Barnett: Phys. Rev. A, **39** 1665 (1989).

[6]   J.H. Shapiro, S.R. Shepard, N.C. Wong: Phys. Rev. Lett., **62** 2377 (1989).

[7]   M. Charbit, C. Bendjaballah, C. Helstrom: I.E.E.E., **IT–35** 1131 (1989).

[8]   W. Schleich, R.J. Horowicz, S. Varro: *Quantum Optics V*, Eds. J.D. Harvey and D.F. Walls, (Springer–Verlag Proceedings in Physics, 1989) **41**, pp. 133–142.

[9]   J.A. Vaccaro, D.T. Pegg: Optics Comm., **86** 529 (1989).

[10]   J.H. Shapiro, S.R. Shepard, N.C. Wong: *Coherence and Quantum Optics VI*, Eds. J.H. Eberly, L. Mandel and E.Wolf (Plenum Press, New York 1990).

[11]   H. le Pas de Sécheval, C. Bendjaballah: *Quantum Aspects of Optical Communications*, Eds. C. Bendjaballah, O. Hirota and S. Reynaud, (Springer–Verlag Lecture Notes in Physics 1991) **378**.

[12]   M.J.W. Hall: Quantum Optics, **3** 7 (1991).

[13]   J.P. Dowling: Optics Comm., **86** 119 (1991).

[14]   A. Vourdas: Phys. Rev., **A 45** 1943 (1992).

[15]   J.W. Noh, A. Fougères, L. Mandel: Phys. Rev. A, **45** 424 (1992).

[16]   S. Abe: Phys. Lett. A, **166** 163 (1992).

[17]   V.B. Braginsky, F. Ya. Khalili: Phys. Lett. A, **175** 85 (1993).

192   Introduction to Photon Communication

[18]   V.P. Belavkin, C. Bendjaballah: Quantum Optics, **6** 169 (1994).
[19]   K.R.W. Jones: Physica Scripta, *to appear*, (1994).
[20]   S.L. Braustein: Phys. Rev., **A 49** 69 (1994).

Conclusion

In this Introduction to quantum communication, we insisted on two points.

It is first stressed that in general the problem of optimal quantum measurement subject to some criterion, cannot be reduced to a problem of a classical statistical reception.

As a second point, it is emphazised that the solution to a problem of quantum detection depends strongly on the minimization of a specified average cost, whereas the maximization of the mutual information transfer depends on the given input states, leaving the parameter that characterizes the redundency, free to vary. These two problems are defined in different contexts and generally have different solutions. In this essay, no attempt was made to establish a correspondence between these approaches.

The two points must however be more elaborated.
First, in a few situations there is some analogy between the quantum and the semi–classical treatments and useful insights can indeed be gained from the classical methods.
Secondly, in classical theory, the maximization of the mutual information associated with hypothesis testing results in determining the threshold of the test, derived from the cost matrix.

Thus, the rate distorsion which can be seen as the minimum quantity of information transfer subject to a given error for making a decision, is a special case of interest. The results for specific cases have been presented together with some elements of a quantum theory. A more detailed analysis certainly warrants further research effort.

Lecture Notes in Physics

For information about Vols. 1–415
please contact your bookseller or Springer-Verlag

Vol. 416: B. Baschek, G. Klare, J. Lequeux (Eds.), New Aspects of Magellanic Cloud Research. Proceedings, 1992. XIII, 494 pages. 1993.

Vol. 417: K. Goeke P. Kroll, H.-R. Petry (Eds.), Quark Cluster Dynamics. Proceedings, 1992. XI, 297 pages. 1993.

Vol. 418: J. van Paradijs, H. M. Maitzen (Eds.), Galactic High-Energy Astrophysics. XIII, 293 pages. 1993.

Vol. 419: K. H. Ploog, L. Tapfer (Eds.), Physics and Technology of Semiconductor Quantum Devices. Proceedings, 1992. VIII, 212 pages. 1993.

Vol. 420: F. Ehlotzky (Ed.), Fundamentals of Quantum Optics III. Proceedings, 1993. XII, 346 pages. 1993.

Vol. 421: H.-J. Röser, K. Meisenheimer (Eds.), Jets in Extragalactic Radio Sources. XX, 301 pages. 1993.

Vol. 422: L. Päivärinta, E. Somersalo (Eds.), Inverse Problems in Mathematical Physics. Proceedings, 1992. XVIII, 256 pages. 1993.

Vol. 423: F. J. Chinea, L. M. González-Romero (Eds.), Rotating Objects and Relativistic Physics. Proceedings, 1992. XII, 304 pages. 1993.

Vol. 424: G. F. Helminck (Ed.), Geometric and Quantum Aspects of Integrable Systems. Proceedings, 1992. IX, 224 pages. 1993.

Vol. 425: M. Dienes, M. Month, B. Strasser, S. Turner (Eds.), Frontiers of Particle Beams: Factories with e+ e⁻ Rings. Proceedings, 1992. IX, 414 pages. 1994.

Vol. 426: L. Mathelitsch, W. Plessas (Eds.), Substructures of Matter as Revealed with Electroweak Probes. Proceedings, 1993. XIV, 441 pages. 1994

Vol. 427: H. V. von Geramb (Ed.), Quantum Inversion Theory and Applications. Proceedings, 1993. VIII, 481 pages. 1994.

Vol. 428: U. G. Jørgensen (Ed.), Molecules in the Stellar Environment. Proceedings, 1993. VIII, 440 pages. 1994.

Vol. 429: J. L. Sanz, E. Martínez-González, L. Cayón (Eds.), Present and Future of the Cosmic Microwave Background. Proceedings, 1993. VIII, 233 pages. 1994.

Vol. 430: V. G. Gurzadyan, D. Pfenniger (Eds.), Ergodic Concepts in Stellar Dynamics. Proceedings, 1993. XVI, 302 pages. 1994.

Vol. 431: T. P. Ray, S. Beckwith (Eds.), Star Formation and Techniques in Infrared and mm-Wave Astronomy. Proceedings, 1992. XIV, 314 pages. 1994.

Vol. 432: G. Belvedere, M. Rodonò, G. M. Simnett (Eds.), Advances in Solar Physics. Proceedings, 1993. XVII, 335 pages. 1994.

Vol. 433: G. Contopoulos, N. Spyrou, L. Vlahos (Eds.), Galactic Dynamics and N-Body Simulations. Proceedings, 1993. XIV, 417 pages. 1994.

Vol. 434: J. Ehlers, H. Friedrich (Eds.), Canonical Gravity: From Classical to Quantum. Proceedings, 1993. X, 267 pages. 1994.

Vol. 435: E. Maruyama, H. Watanabe (Eds.), Physics and Industry. Proceedings, 1993. VII, 108 pages. 1994.

Vol. 436: A. Alekseev, A. Hietamäki, K. Huitu, A. Morozov, A. Niemi (Eds.), Integrable Models and Strings. Proceedings, 1993. VII, 280 pages. 1994.

Vol. 437: K. K. Bardhan, B. K. Chakrabarti, A. Hansen (Eds.), Non-Linearity and Breakdown in Soft Condensed Matter. Proceedings, 1993. XI, 340 pages. 1994.

Vol. 438: A. Pękalski (Ed.), Diffusion Processes: Experiment, Theory, Simulations. Proceedings, 1994. VIII, 312 pages. 1994.

Vol. 439: T. L. Wilson, K. J. Johnston (Eds.), The Structure and Content of Molecular Clouds. 25 Years of Molecular Radioastronomy. Proceedings, 1993. XIII, 308 pages. 1994.

Vol. 440: H. Latal, W. Schweiger (Eds.), Matter Under Extreme Conditions. Proceedings, 1994. IX, 243 pages. 1994.

Vol. 441: J. M. Arias, M. I. Gallardo, M. Lozano (Eds.), Response of the Nuclear System to External Forces. Proceedings, 1994, VIII. 293 pages. 1995.

Vol. 442: P. A. Bois, E. Dériat, R. Gatignol, A. Rigolot (Eds.), Asymptotic Modelling in Fluid Mechanics. Proceedings, 1994. XII, 307 pages. 1995.

Vol. 443: D. Koester, K. Werner (Eds.), White Dwarfs. Proceedings, 1994. XII, 348 pages. 1995.

Vol. 444: A. O. Benz, A. Krüger (Eds.), Coronal Magnetic Energy Releases. Proceedings, 1994. X, 293 pages. 1995.

Vol. 445: J. Brey, J. Marro, J. M. Rubí, M. San Miguel (Eds.), 25 Years of Non-Equilibrium Statistical Mechanics. Proceedings, 1994.

Vol. 446: V. Rivasseau (Ed.), Constructive Physics. Results in Field Theory, Statistical Mechanics and Condensed Matter Physics. Proceedings, 1994. X, 337 pages. 1995.

Vol. 447: G. Aktaş, C. Saçlıoğlu, M. Serdaroğlu (Eds.), Strings and Symmetries. Proceedings, 1994. XIV, 389 pages. 1995.

Vol. 448: P. L. Garrido, J. Marro (Eds.), Third Granada Lectures in Computational Physics. Proceedings, 1994. XIV, 346 pages. 1995.

Vol. 449: J. Buckmaster, T. Takeno (Eds.), Modeling in Combustion Science. Proceedings, 1994. X, 369 pages. 1995.

Vol. 450: M. F. Shlesinger, G. M. Zaslavsky, U. Frisch (Eds.), Lévy Flights and Related Topics in Physics. Proceedigs, 1994. XIV, 347 pages. 1995.

Vol. 452: A. M. Bernstein, B. R. Holstein (Eds.), Chiral Dynamics: Theory and Experiment. Proceedings, 1994. VIII, 351 pages. 1995.

Vol. 453: S. M. Deshpande, S. S. Desai, R. Narasimha (Eds.), Fourteenth International Conference on Numerical Methods in Fluid Dynamics. Proceedings, 1994. XIII, 588 pages. 1995.

Vol. 454: J. Greiner, H. W. Duerbeck, R. E. Gershberg (Eds.), Flares and Flashes, Germany 1994. XXII, 477 pages. 1995.

New Series m: Monographs

Vol. m 1: H. Hora, Plasmas at High Temperature and Density. VIII, 442 pages. 1991.

Vol. m 2: P. Busch, P. J. Lahti, P. Mittelstaedt, The Quantum Theory of Measurement. XIII, 165 pages. 1991.

Vol. m 3: A. Heck, J. M. Perdang (Eds.), Applying Fractals in Astronomy. IX, 210 pages. 1991.

Vol. m 4: R. K. Zeytounian, Mécanique des fluides fondamentale. XV, 615 pages, 1991.

Vol. m 5: R. K. Zeytounian, Meteorological Fluid Dynamics. XI, 346 pages. 1991.

Vol. m 6: N. M. J. Woodhouse, Special Relativity. VIII, 86 pages. 1992.

Vol. m 7: G. Morandi, The Role of Topology in Classical and Quantum Physics. XIII, 239 pages. 1992.

Vol. m 8: D. Funaro, Polynomial Approximation of Differential Equations. X, 305 pages. 1992.

Vol. m 9: M. Namiki, Stochastic Quantization. X, 217 pages. 1992.

Vol. m 10: J. Hoppe, Lectures on Integrable Systems. VII, 111 pages. 1992.

Vol. m 11: A. D. Yaghjian, Relativistic Dynamics of a Charged Sphere. XII, 115 pages. 1992.

Vol. m 12: G. Esposito, Quantum Gravity, Quantum Cosmology and Lorentzian Geometries. Second Corrected and Enlarged Edition. XVIII, 349 pages. 1994.

Vol. m 13: M. Klein, A. Knauf, Classical Planar Scattering by Coulombic Potentials. V, 142 pages. 1992.

Vol. m 14: A. Lerda, Anyons. XI, 138 pages. 1992.

Vol. m 15: N. Peters, B. Rogg (Eds.), Reduced Kinetic Mechanisms for Applications in Combustion Systems. X, 360 pages. 1993.

Vol. m 16: P. Christe, M. Henkel, Introduction to Conformal Invariance and Its Applications to Critical Phenomena. XV, 260 pages. 1993.

Vol. m 17: M. Schoen, Computer Simulation of Condensed Phases in Complex Geometries. X, 136 pages. 1993.

Vol. m 18: H. Carmichael, An Open Systems Approach to Quantum Optics. X, 179 pages. 1993.

Vol. m 19: S. D. Bogan, M. K. Hinders, Interface Effects in Elastic Wave Scattering. XII, 182 pages. 1994.

Vol. m 20: E. Abdalla, M. C. B. Abdalla, D. Dalmazi, A. Zadra, 2D-Gravity in Non-Critical Strings. IX, 319 pages. 1994.

Vol. m 21: G. P. Berman, E. N. Bulgakov, D. D. Holm, Crossover-Time in Quantum Boson and Spin Systems. XI, 268 pages. 1994.

Vol. m 22: M.-O. Hongler, Chaotic and Stochastic Behaviour in Automatic Production Lines. V, 85 pages. 1994.

Vol. m 23: V. S. Viswanath, G. Müller, The Recursion Method. X, 259 pages. 1994.

Vol. m 24: A. Ern, V. Giovangigli, Multicomponent Transport Algorithms. XIV, 427 pages. 1994.

Vol. m 25: A. V. Bogdanov, G. V. Dubrovskiy, M. P. Krutikov, D. V. Kulginov, V. M. Strelchenya, Interaction of Gases with Surfaces. XIV, 132 pages. 1995.

Vol. m 26: M. Dineykhan, G. V. Efimov, G. Ganbold, S. N. Nedelko, Oscillator Representation in Quantum Physics. IX, 279 pages. 1995.

Vol. m 27: J. T. Ottesen, Infinite Dimensional Groups and Algebras in Quantum Physics. IX, 218 pages. 1995.

Vol. m 28: O. Piguet, S. P. Sorella, Algebraic Renormalization. IX, 134 pages. 1995.

Vol. m 29: C. Bendjaballah, Introduction to Photon Communication. VII, 193 pages. 1995.

Vol. m 30: A. J. Greer, W. J. Kossler, Low Magnetic Fields in Anisotropic Superconductors. VII, 161 pages. 1995.

Vol. m 31: P. Busch, M. Grabowski, P. J. Lahti, Operational Quantum Physics. XI, 230 pages. 1995.

Springer-Verlag
and the Environment

We at Springer-Verlag firmly believe that an international science publisher has a special obligation to the environment, and our corporate policies consistently reflect this conviction.

We also expect our business partners – paper mills, printers, packaging manufacturers, etc. – to commit themselves to using environmentally friendly materials and production processes.

The paper in this book is made from low- or no-chlorine pulp and is acid free, in conformance with international standards for paper permanency.